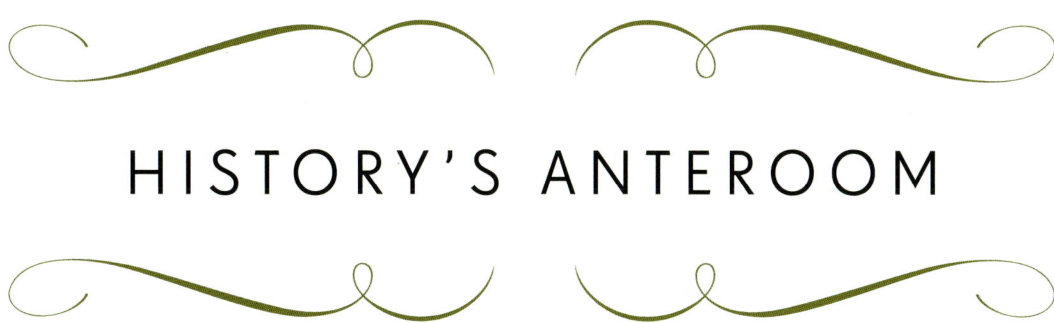

HISTORY'S ANTEROOM

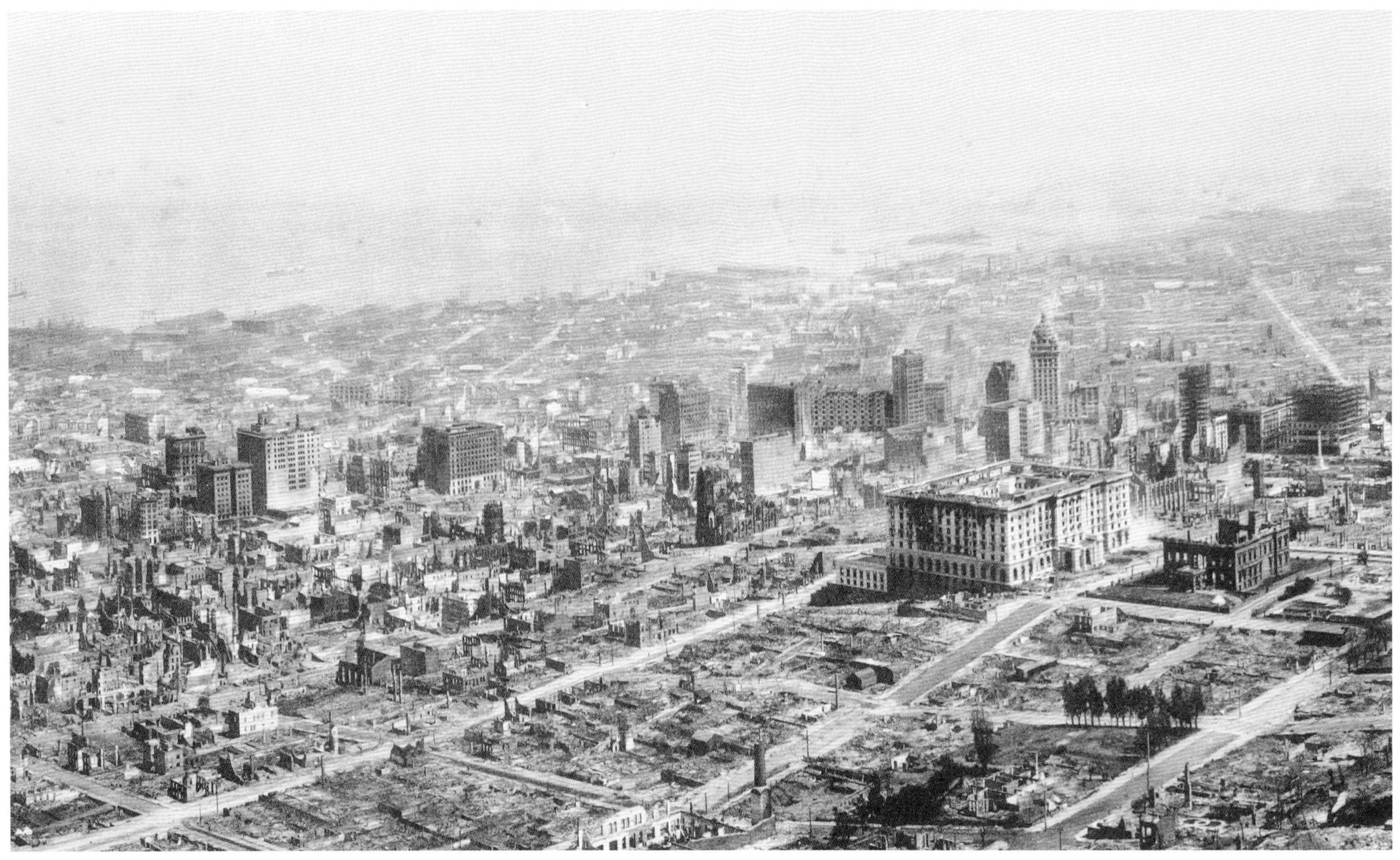

GEO. R. LAWRENCE CO.

RUINS OF SAN FRANCISCO NOB HILL IN FOREGROUND FROM LAWRENCE CAPTIVE AIRSHIP, 1,500 FEET ELEVATION, MAY 29, 1906, PHOTOLITHOGRAPH

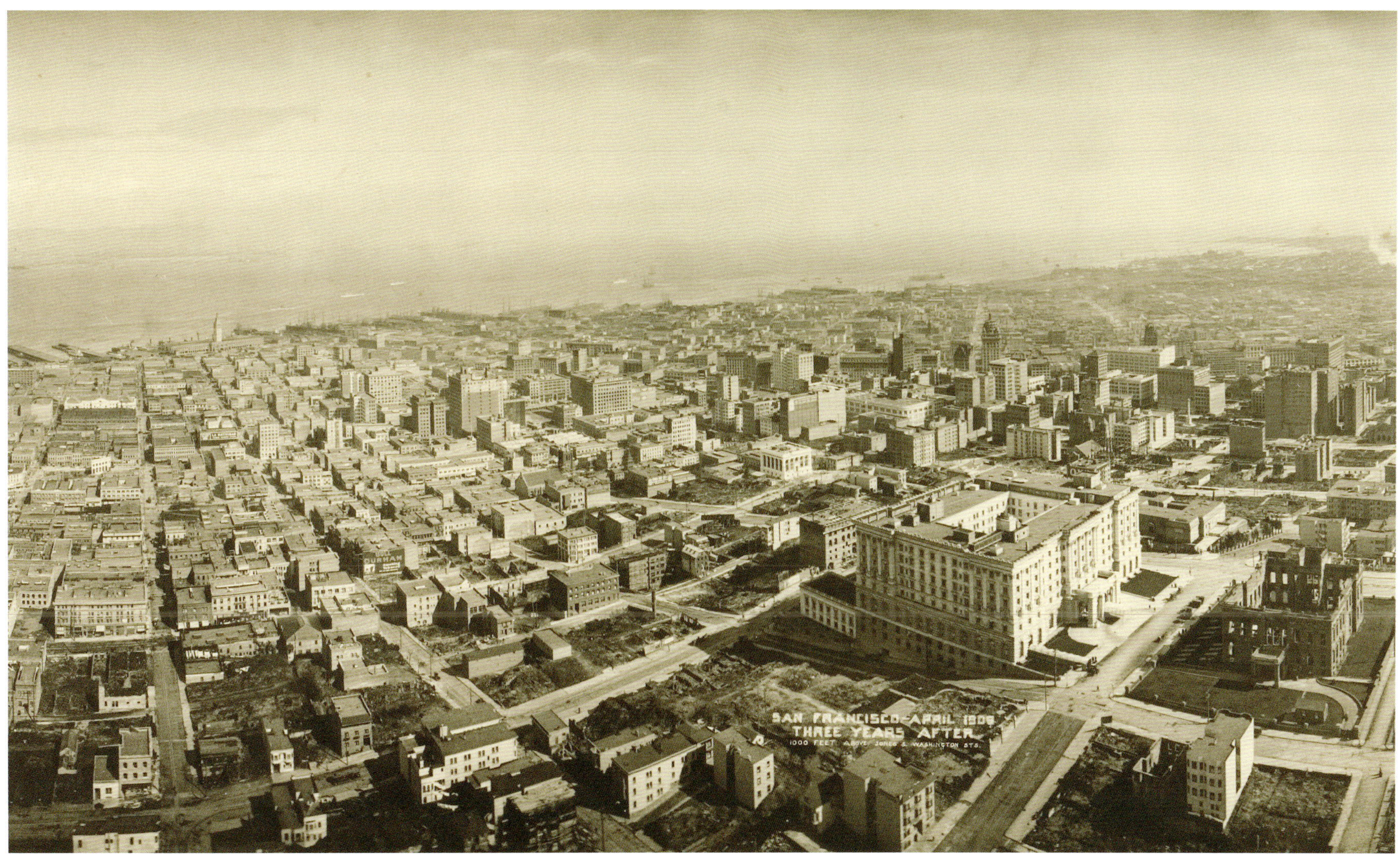

R. J. WATERS AERIAL PHOTOGRAPH CO.

SAN FRANCISCO, THREE YEARS AFTER, 1,000 FEET ABOVE JONES & WASHINGTON STS., APRIL 1909, PHOTOLITHOGRAPH

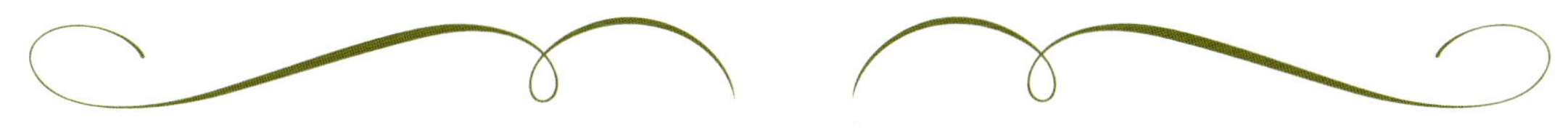

HISTORY'S ANTEROOM

Photography in San Francisco 1906–1909

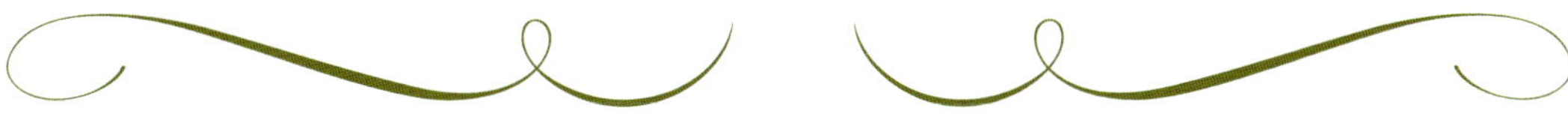

RODGER C. BIRT

PHOTOGRAPHIC DESCRIPTIONS BY

MARVIN R. NATHAN

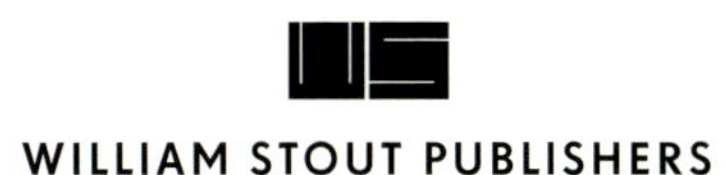

WILLIAM STOUT PUBLISHERS

WILLIAM STOUT PUBLISHERS

1326-1328 South 51st Street
Richmond, CA 94804
www.stoutpublishers.com

Library of Congress Control Number: 2011924769
ISBN: 978-0-9819667-5-5

The authors would like to thank the following individuals whose help was
instrumental in making this book possible: Professor Arthur Chandler,
Director, Alvin Fine Chair of San Francisco Studies at San Francisco State
University; Susan Goldstein, City Archivist, San Francisco History Center,
Book Arts and Special Collections, San Francisco Public Library; Ian
Andrews and Tamara Bock, researchers; Lorna Hill, manuscript preparation;
Jenny Thompson, editorial assistance; Laura Ciapponi, photography; the
individual collectors who wished to remain anonymous but shared their
private collections with us; and the librarians at these institutions: the
Huntington Library, the National Archives, the Society of California
Pioneers, the Visual Studies Workshop, and the San Francisco History
Center, San Francisco Public Library.

The authors and publisher gratefully acknowledge the generous funding
this book received from: The Alvin Fine Chair of San Francisco Studies
at San Francisco State University and The Friends of the San Francisco
Public Library

Cover and book design by Debbie Berne / www.debbiebernedesign.com

Printed in China

To D.E.R. and L.R.-B.
R.C.B.

To G.S.N.
and to my father, whose
birth certificate perished in 1906
M.R.N.

PHOTOGRAPHY IN SAN FRANCISCO 1906–1909: THE HISTORICAL MOMENT

Photography is the anteroom of history. Siegfried Kracauer (1889–1966)

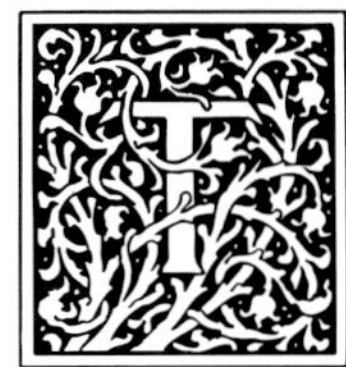
he inherently singular characteristic of a photograph, that quality unique to the light-written image, is its authoritative statement of fact(s): the photographer saw that which you and I now see, the visible traces—tones, textures, geometries—of some other, previous moment. But the viewer is more than merely a passive recipient of information. When we look at a photograph, we consciously call on memory and thereby provide the essential connection between what was and what is. Through the conspiracy of maker and consumer, past becomes present and time avails not. A photograph, while fixing a slice of time (past), also represents, for the viewer, a moment *in time* (present). Because of this special relationship to time, the German critic of culture, Siegfried Kracauer (1889–1966), understood that still photography (and its kinetic corollary, cinema) could become a tool of considerable value for students of history. Kracauer believed that photographs and films provided historians opportunities to "enter" other eras in very literal ways and actually "be among" the participants of those times. While attempting to understand the process of writing history, Kracauer says he came to realize that "the historian has traits of the photographer and historical reality resembles camera reality." This assessment of photographs and their role in historical analysis was part of his interest in historiography. Kracauer realized that photographs are not just "windows" into the past: more importantly, he observed, they are history's "redemption." In his analysis, time travel is no fantasy. It is a fact—the "actuality" of every photograph.[1]

How photographs actualize the palpable reality of a period informs this analysis of a group of selected photographers and their work in San Francisco between 1906 and 1909. In the following essay I have designated the period under investigation the "Event." Comprised of discrete, chronological divisions, the Event began with an earthquake on April 18, 1906, and ended on October 23, 1909, the final day of a carnival named the Portola Festival which celebrated the rebuilt city. The four phases of the Event were earthquake, fire, immediate relief, and the long-term rehabilitation signaled by reconstruction. The photographs made during the three phases following the earthquake are simultaneously windows, mirrors, and, most fabulously, portals into those other times. They illuminate not only the earthquake's effects but also the survivors' attempts to take stock of the destruction, the city's efforts to reconstitute itself, and our own ability to understand the Event from the photographic archive we have inherited.

First introduced into the United States in 1840, photography had replaced all other visual media as the preferred means to provide information pictorially by the early 1900s. Photographs supplanted engravings, lithographs, woodcuts, and paintings, and few events in the nation's life escaped the photographer's scrutiny. Therefore, when a powerful earthquake, followed by a devastating fire, struck San Francisco and much of the rest of northern California, photographers gave full attention to the temblor and its effects, both immediately after the tragedy and in the years that followed. The result of their efforts was a vast archive of visual data.

The photographers who took part in compiling this visual record of the Event form a variegated group. They were commercial photographers, U.S. government-employed photographers, art photographers, and amateur photographers. Their responses to the visual data they confronted were as varied as their individual relationships to the medium. Over the span of the hundred years since their prints were first made, their photographs, for the most part, have languished in archives or in private collections, and only a small number of them have been available for viewing and scholarly examination. Now, in many cases for the first time, these photographers and their work are brought together. Henry W. Chadwick and D. H. Wulzen join the ranks of previously identified photographers of the Event. (Among the latter there are varying degrees of public awareness and scholarly knowledge of their extant prints.) Arnold Genthe (1869–1942) has been the subject of a good deal of attention; Willard E. Worden (1868–1946) has received scant interest; the prints made by unidentified photographers working for the corporate entities, the Pillsbury Picture Company and R. J. Waters Company, have been featured in recent auctions; Louis J. Stellman (1877–1961) has been written about almost entirely as a photographer of Chinatown and a follower of Genthe; James D. Givens and Edgar A. Cohen (1859–1939) await their chroniclers; and the vernacular photographer, Crittenden Van Wyck, has never had his work associated with photographers such as those listed above.[2]

I

Although the north central California earthquake of 1906 remains lodged in popular imagination a century later as the "Great San Francisco Earthquake," the temblor's destructive force wreaked havoc upon regions of California far to the north and south of the San Francisco Bay metropolis. West of the city, under the Pacific Ocean, the earth's annihilating forces were first unleashed. Ten miles to the east, suburban towns such as Berkeley and Oakland were shaken but emerged relatively unscathed. In Berkeley no deaths were reported, and there was only minimal property damage. In Oakland five people died when a wall of the Empire Theatre at Twelfth and Washington Streets collapsed into a neighboring hotel. Physically, however, Oakland remained intact for the most part, and in the days and weeks to follow it would become a vital center for relief and the logistics of reconstruction. Further east, beyond the East Bay hills in sparsely populated, agriculture-dominated regions, the earthquake was little more than a momentary nuisance. Along the Pacific coast, however, the story was dramatically different. Here lay the fault zone, and from Eureka, two hundred fifty miles north of San Francisco, to Salinas, ninety miles to the southeast, the loss of life and the destruction of property combined to make the event of Wednesday, April 18, 1906, one of the nation's deadliest and most costly (*fig. 1*).[3]

At 5:11 A.M. the movement of the earth's tectonic plates under the ocean floor was first felt in Eureka; the crew aboard the steamer "Argo" actually saw a bolt of force strike the coast. At Eureka the United States Weather Bureau reported that the shaking lasted forty-seven seconds and was the most severe recorded there since the station had opened in 1887. Eureka and the smaller towns of Arcata and Trinidad nearby suffered massive damage and large losses of life. (*Daily Humboldt Standard,* April 18). The state of panic and dread caused by the early morning horror was reinforced later in the day when the *Standard* reported (erroneously) that "Oakland and Berkeley were destroyed with over ten thousand dead." A scant thirty seconds after the earthquake first began to move the ground beneath Eureka, the town of Santa Rosa, one hundred and eighty miles to the south, was jolted.

Relative to size and population this small town of less than six thousand residents was the hardest hit of all the earthquake-stricken communities. Forty-six people died there. Virtually every building was leveled, and those that survived the shaking would be consumed by fire shortly after. As the first wisps of smoke began to rise into Santa Rosa's dust-filled air, the earthquake slammed into Sonoma and Marin County towns.[4]

Racing along its ancient, geophysical path—the San Andreas fault line—the temblor's force hurtled through Guerneville where it entombed three "graveyard shift" workers at the Great Eastern quicksilver mine. The impact was so tremendous when it struck in Point Reyes, it was able to knock a massive steam locomotive and its three cars off the tracks. The train was scheduled to depart at 5:15 A.M. At Fort Ross the almost century-old Russian Orthodox Church lurched from its foundation and its steeple crashed down beside the newly exposed nave. Tomales, the most devastated of all the Marin County towns, along with Healdsburg, Olema, and Sebastopol, simultaneously shuddered and then began to break apart. The shaking continued for half a minute. In that thirty-second interval the shock—which was first recorded at Eureka and actually seen by the crew of the "Argo"—burst upon San Francisco: it was twelve minutes and fifteen seconds past five o'clock in the morning.[5]

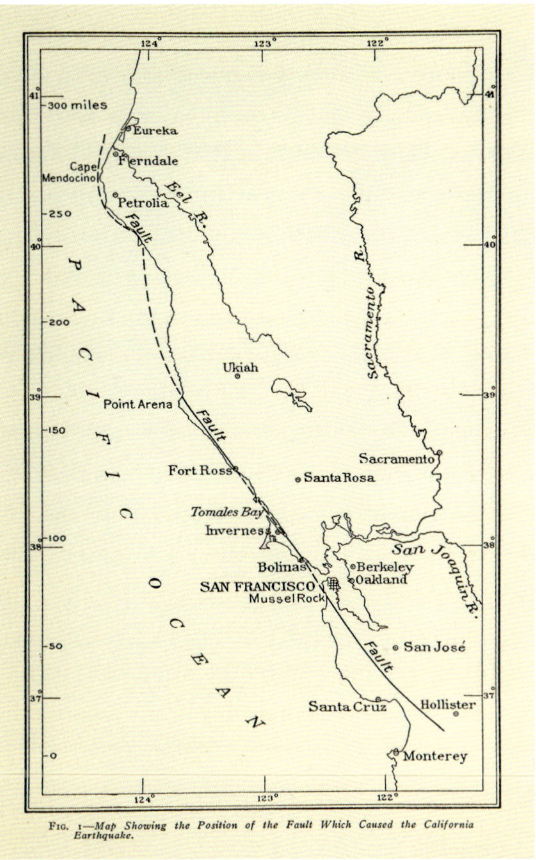

fig. 1. Map showing the area affected by April 1906 earthquake (private collection)

Across the peninsula south of San Francisco and into the South Bay the earth's unleashed forces had hurried. San Mateo, Redwood City, Millbrae, Burlingame, and Mountain View were all pummeled a few seconds after the city was hit. Some places, notably Los Gatos, were spared.

Palo Alto was the most severely impacted of all South Bay locations, and the neighboring campus of Stanford University was the site of spectacular destruction. Happily, the loss of life was slight: only two deaths were recorded at Stanford. But William James, the noted psychologist and brother of novelist Henry James, who was just completing his stay at Stanford as a visiting professor, noted that the survivors appeared to be in a state of "trance-like animation." This was among James's first observations of human behavior after great trauma that eventually would lead to his theory that these men and women were the victims of what psychologists now label post traumatic stress disorder. But even before James and his frightened and dazed fellow victims could escape the crumbling buildings, San Jose, ten miles to the south, was feeling the earthquake's force.[6]

San Jose, William Bronson writes, "was San Francisco in miniature that day." Destruction rained on the new Hall of Justice, the town's churches, and several of its largest business blocks. Throughout the city frame houses were tossed off their foundations, and a daylong fire began to rage within minutes of the first sickening movement of the earth. When there was time to count the homeless, they numbered in excess of eight thousand, though only nineteen had died. A few miles to the north of downtown one of the saddest of all the stories of that morning was unfolding at the Agnew State Asylum. Bedlam reigned at Agnew's all that day and for most of the next week. The death toll was enormous; when the final census was taken, one hundred and seven patients had been killed as well as twelve attendants and doctors. Students and teachers from nearby Santa Clara University rushed to help in the rescue efforts. They pulled the wounded out of the ruins and attempted to comfort the panic-stricken inmates. The more violent among the ill were strapped to trees with sheets and blankets to prevent them from doing harm to themselves or their fellow patients.[7]

At 5:15 A.M., three minutes after the earthquake's force was first recorded in San Francisco and four minutes after the agony had begun in Eureka and elsewhere in Humboldt County, the southernmost edges of the arc of destruction began to experience violent shaking. The temblor was unleashing its final fury. In Salinas, Los Banos, Hanford, Sanger, Boulder Creek, Gilroy, and all the way to Visalia, chimneys collapsed, buildings lurched off their foundations, and rudely awakened dreamers clambered into the open. Here loss of life was small but tragic nevertheless. At the Del Monte Hotel in Monterey the manager and a visiting couple from Oregon on their honeymoon were killed. In Hollister a man was reported to have committed suicide when he saw his wife's dead body.[8]

The end of the geological phase of the Event was reported at a few seconds after 5:15 A.M. on the grounds of the Spanish Mission San Juan Bautista. Here the mission church and adjoining buildings all suffered grave damage. Completed in 1812, the complex of buildings had withstood every previous temblor. But on that April morning the ninety-year-old structures were so severely damaged that reconstruction efforts would last a decade. This was the end of the earthquake's destructive course. Reports of only mild shaking were recorded south of Mission San Juan Bautista. Almost four hours would elapse before the Los Angeles and San Diego telegraph stations first sent word of the disaster out across the rest of the country and beyond. By that time, for the survivors along the almost three-hundred-mile path of destruction, the first challenges, set in motion by the catastrophic event, were well underway.[9]

II

In San Francisco, among the thousands rushing from the shaking buildings were dozens of photographers: professional and commercial photographers, magazine and newspaper photographers, skilled amateurs, and "snapshot" enthusiasts. Within hours of the first phase of the Event, military photographers began the official photographic documentation their commanding officers had ordered. Edgar A. Cohen, a respected Bay Area amateur photographer, noted that wherever the earthquake had struck, photographers had made pictures. Most photographers lived within the zone of destruction itself or came from nearby, but some had traveled great distances to ensure having an opportunity to secure images of the devastation. The local photographers were able to capture images of the fire itself as it raged for three days. These prints of the conflagration proved to be in greatest demand initially. Everybody wanted "smoke pictures," Cohen remarked. Images of near encounters with primal, powerful forces beyond human control, the photographs would fascinate a global audience.[10]

The Pillsbury Picture Company, an Oakland-based firm, issued a list of more than four hundred photographs of the "fire and ruins" and sold prints to individual buyers and to publications. In May, Pillsbury published a set of prints that included three photographs of afternoon views of the fire: one (pl. 11) was taken north of Market Street; a second (pl. 12) from the northeast corner of Market and Montgomery showing the burning of the Palace Hotel; and the third (pl. 10) taken on Mission Street near New Montgomery. The three Pillsbury prints were among the large number included in an illustrated essay about the earthquake and fire in the May 1906 issue of the *New San Francisco Magazine,* and one (pl. 12) appeared as an illustration in Charles Keeler's eyewitness account published in the autumn.[11]

Arnold Genthe, in 1906 a studio photographer of San Francisco's wealthy elite who earlier had gained fame as the photographer of Chinatown, did not begin to photograph until late in the morning of the first day. He first needed to borrow film, a camera, and lenses from George Kahn, a camera supplier he knew, since his own equipment had been destroyed in the initial shock and subsequent collapse of his apartment building on Sutter Street. In contrast, Willard E. Worden, an established commercial and view photographer, began working early in the morning, starting out from his lower Nob Hill home within minutes of the first shaking. From Nob Hill, Worden worked his way toward the fires now growing in intensity South of Market Street in the working class neighborhood known as—because it was south of the cable car tracks—"South of the Slot" (pl. 8).

After making his exposures of the fires South of Market Street, Worden proceeded north into the blocks surrounding Union Square Park and continued to photograph. One of the resulting prints shows the early stages of the Call Building fire. The city's tallest building at the time and home of the newspaper, the *San*

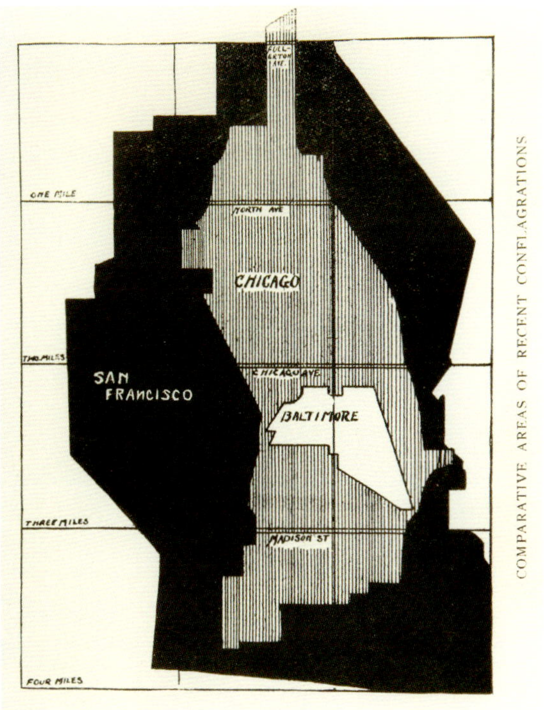

fig. 2. The three largest urban fires of the period (private collection)

Francisco Call, the tower burned from the top down after wind-borne embers from the raging South of Market fire gained access to the ornate, Beaux Arts dome (pl. 6). At this vantage point Worden was looking south from O'Farrell Street towards the southeast intersection of Market and Third Streets where the Call Building stood. The spectacle of the burning edifice captivated Worden and he photographed it from a second

vantage point several blocks south and nearer to Market Street (pl. 9). (Later, a Pillsbury photographer would photograph from this same location.) By now, just past noon, the upper floors of the Call Building were entirely engulfed in flames, but the Palace Hotel had not yet been invaded by the fire.[12]

After taking several photographs along Market Street, Worden entered the inferno blazing in South of the Slot. He took dramatic close-up views of burning buildings and the crowds of onlookers. Soon photographer and crowd would be driven away as the fire lashed out in every direction and eventually jumped north across Market Street, destroying everything in its path. It was now early afternoon and the fire had already seared itself into American history; in tandem with the earthquake it would become the nation's largest urban disaster (*fig. 2*).

By midday, all of downtown San Francisco on both sides of Market Street was in flames and the fire was consuming everything in its path, east to the bay, north towards the Golden Gate, and south and west, into the city's industrial areas and nearing the Mission Dolores, the city's oldest structure. As the downtown temperature rose to an estimated two thousand seven hundred degrees Fahrenheit, crowds of onlookers were forced to retreat to what they thought was the safety of the higher elevations on the city's several hills. Around this time Arnold Genthe began his examination of the event.[13]

Since Genthe had first been shaken awake, eight hours had elapsed. Now the horrible enormity of the situation was evident. For several hours the fire had seized everyone's attention, but at 10:15 A.M. there came an aftershock of significant force. Jolts of varying intensity would continue all day as the fire grew in ferocity, an implacable foe seemingly beyond human control. Now that the survivors were outdoors, in streets, parks, and cemeteries—anywhere where only the smoke-filled sky was above their heads—all seemed to accept the movement of the ground beneath their feet with an almost stoic calm. Transfixed, they gathered to watch the fire. Worden had earlier caught crowds gazing at the Call Building slowly burning floor by floor. Genthe's photographs of the spectacle of fire, taken six hours later, capture the immensity of the tragedy and how mesmerizing it was for the temblor survivors. In his autobiography, Genthe described the published view (pl. 13) as a "pictorially effective composition [showing] the results of the earthquake, the beginning of the fire, and the attitude of the people."[14]

While Genthe and Worden were roving through downtown, Carleton E. Watkins (1829–1916), one of the previous generation's most brilliant photographers, was being led from

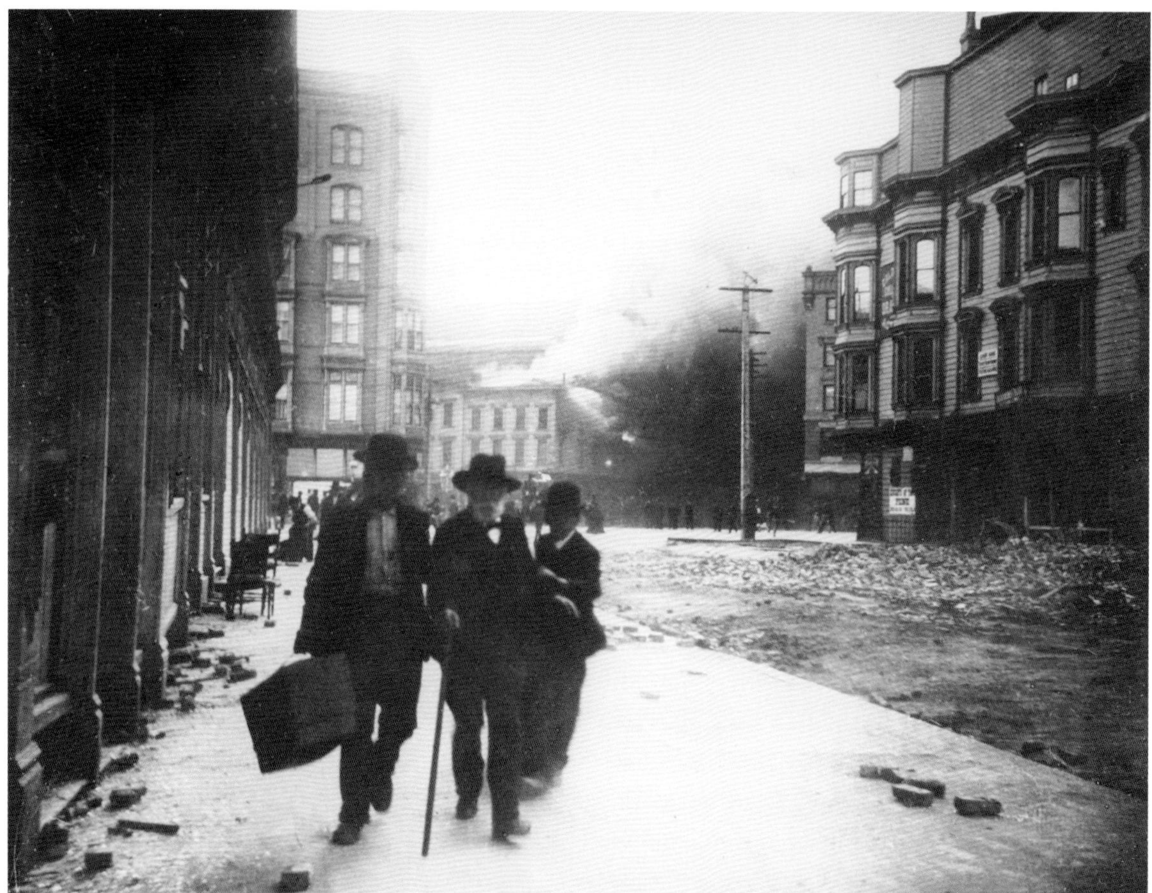

fig. 3. Carleton Watkins led through the burning city (Society of California Pioneers)

his studio as the South of Market fire reached within a hundred yards of the building he had occupied for years; the flames were threatening to destroy his collection of views of the entire western United States (*fig. 3*). Ironically, a now blind Watkins could see nothing of what was transpiring all around him. His son, Collis, and a friend, Charles Turill, are seen leading him out of the building into the safety of the street; forced to leave behind thousands of his prints and negatives, Watkins ultimately lost everything. His forced evacuation was the first step of his decline into senile dementia. Guided to a friend's house, he soon retired to his ranch in Sonoma where he would spend the next four years of his life. In 1910 his family committed him to the Napa State Hospital where he would die six years later. But on that April night in 1906 Watkins slept outdoors with Collis and Turill, while hundreds of others camped out nearby.[15]

As Watkins and his companions tried to rest, an eerie calm settled over the city. The only sounds were the hiss and whisper of the fires, the trudging of human feet, and the rumble of carriages and wagons, some drawn by horses, others by men. All were heading away from the conflagration towards the ocean and the open spaces in the city's western precincts. Behind the evacuating masses, in the burning streets, army munition experts were dynamiting entire blocks of buildings. The hope was to create firebreaks before the fire could reach them, feed on them, and grow in intensity. "Destroy to save" became the order, and indeed San Francisco looked and sounded like a hellish battle zone. Thursday was a desperate time, and Friday just as bad. But by Saturday the flames had been checked; the Van Ness Avenue line had held the fire back during all this time, and the army and navy helped maintain order. On Saturday it began to rain, a vigorous, merciful rain that continued throughout the day. The second phase of the Event—the fire—ended here, and a longer, and perhaps more difficult time was about to begin.[16]

First, thirst and famine had to be dealt with. On Thursday, the second day, thousands went without water. By Friday almost everyone was facing a third day without food. Genthe's breakfast on Wednesday, taken only hours after the first shaking, would be his last meal until Saturday, but by Sunday the coordinated efforts to bring order and restore a semblance of normal civic life were in place. United States Army and Navy contingents from all over the Bay Area and neighboring states had become involved in every aspect of the Event's unfolding story from the earliest hours of Wednesday morning, April 18, until their withdrawal at the end of June. These units helped to fight the fires, controlled unruly crowds and individuals, established a water and food distribution network, and assisted in the building of the earliest refugee camps and temporary infrastructures. Throughout the earthquake and fire-damaged zone, military medical expertise would be vital in maintaining public health and preventing the outbreak of disease. Brigadier General Frederick Funston served as the commanding officer of the army units from April 18 to April 23, and the officer in charge from April 23 until the units were withdrawn on June 30 was Major General Adolphus W. Greely. Greely's U.S. Navy counterpart was Commander Charles J. Badger. By the time Greely and Badger ended their liaison with civilian jurisdictions and relief agencies, the city had been organized into twelve relief and distribution centers, while beyond San Francisco the providing of aid was efficiently taking place without further military assistance. Photographs documenting the weeks of immediate relief after the devastation of earthquake and fire comprise a second significant body of photographic work surviving from this period.[17]

III

In addition to Worden, Genthe, and the Pillsbury Picture Company photographers, all of whom remained active after the fires had subsided and well into the summer of 1906, a host of other photographers also used the destroyed cityscape as a zone of activity. The most noteworthy among this latter group are Corporal Henry W. Chadwick, assigned to the U.S. Army Signal Corps at the Presidio, James D. Givens, a civilian commercial photographer with a studio at the Presidio and various U.S. Army contracts, and two especially talented amateur photographers, D(ietrich) H(einrich) Wulzen, a pharmacist, and Louis J. Stellman, a newspaper journalist and aspiring poet.[18]

The efficiency with which relief was delivered and the first steps to recovery were taken was only partly due to the presence of the army and navy. Much of the credit has to be given to the surprisingly nonpartisan leadership of city, state, and federal government. Mayor Eugene Schmitz held meetings with various officials throughout Wednesday at a series of temporary city headquarters since the actual center, the seven-year-old City Hall, had been all but brought down by the early morning temblor (pl. 3). Early in the morning on Thursday, Schmitz organized a Citizens Committee comprised of leading figures representing local business and political circles. The Citizens Committee would direct all organizations involved with immediate relief and later, long-term recovery. More than two hundred thousand citizens were homeless, while those whose homes had not been destroyed were required to live outdoors until chimneys and gas fittings could be examined and deemed safe. In consequence, temporary camps sprang into existence all over San Francisco and in nearby towns and cities, especially across the bay in Oakland, where as many as one hundred fifty thousand displaced San Francisco homeless found shelter in makeshift camps.[19]

During the earliest weeks of life in the camps, the temporary structures themselves were precariously fragile: all literally had been put together with any available materials (pl. 31). The distribution of water and food began here; at first sailors and soldiers—commanded to do so by their superior officers—had taken commodities from shops and restaurants in sectors they had been assigned to patrol and doled out what they had found to the needy. By the end of the first week several thousand tons of food and other essential items—tents, blankets, shoes, clothing—had arrived in San Francisco and were distributed at army relief stations.

During this period Henry Chadwick was one of the most active of all the military photographers. And, because of his local family ties, Chadwick continued to photograph even after the completion of his official assignment. He eventually assembled an album of sixty-six prints. Some of the prints in the album were included in the Signal Corps records, while others reflect more personal vignettes.[20]

The prints Chadwick made for the army were delivered to Washington, D.C., where they became a part of the voluminous data the Office of the Secretary of War was gathering on the Event. By the time Chadwick and Givens were working, some families had been reunited and others had had to absorb the full meaning of their private tragedies. During the following days, as these photographers continued to document the devastated city, the transit system was restored line by line, businesses reopened—occasionally at original locations, but more often at new addresses—and the amity and communal fellowship engendered by the tragedy in the first few weeks gave way to the more typical divisions and conflicts between classes and among ethnicities to be found everywhere in large American cities in the early twentieth century. However, Chadwick's photographs celebrate the dispossessed and their attempts to establish community in the temporary camps (pls. 28–30), while Givens concentrates on the order and calm of the devastated city now safely in the hands of the federal forces (pls. 35–39). Givens's photographs were reproduced as part of the Commissary General Office's *June Report on Relief Efforts in San Francisco.* At the end of 1906 he published a comprehensive photographic essay on the early months of the Event in *San Francisco in Ruins,* which featured photographs by Givens and other leading photographers. For Givens, Chadwick, and the photographers represented in *San Francisco in Ruins* the vision was overwhelmingly optimistic.[21]

Since the late 1880s pictorial photography had flourished in the Bay Area. This movement strived to make camera images more painterly and more artistically personalized by the use of muted tones, soft focus, and darkroom manipulation to achieve emotional force and symbolic meaning. In 1892 the California Camera Club was formed, one of the few such organizations in the United States to encourage professional photographers' membership. Shortly after the club's founding, the journal *Camera Craft* was launched, and communities of pictorialist photographers developed in Oakland and Berkeley as well as in San Francisco. In 1901 Bay Area pictorialists held their first International Salon, and the exhibited work attracted critical acclaim. After that event, California Camera Club members were regularly invited to exhibit work in Chicago and New York. By 1906 Oscar Maurer and his pictorialist colleagues, especially Laura Adams Armer, Anne W. Brigman, and William Dassonville, made the Bay Area a center of pictorialism. Maurer and Genthe photographed extensively after the earthquake and fire, and Maurer contributed prints to Charles Keeler's *San Francisco Through Earthquake and Fire.* Armer, Brigman, and Dassonville, however, seem not to have taken the opportunity to photograph the dramatic ruinscape. D. H. Wulzen, the San Francisco–born son of German immigrant parents, was a friend of Arnold Genthe and a respected pharmacist in the city's large ethnic German community. He joined the California Camera Club in the late 1890s. Wulzen made a few views of the burning city but concentrated his attention on the postfire period. His photographs, like some of Chadwick's, show a personal involvement in

quotidian elements of life during recovery. His pharmacy, located on the southeast corner of the intersection of Seventeenth and Castro streets, served as an emergency food distribution center and temporary post office. Photographs of a line of his neighbors waiting to pick up mail (pl. 46) and of his wife preparing a meal in the temporary kitchen in front of their property (pl. 47) not only record his personal connection to events but also reflect the private realities of hundreds of thousands of his fellow citizens. Waiting in long lines and waiting for food and water were everybody's lot, and cooking al fresco continued throughout the summer and into the autumn, ending only after city inspectors determined the safety of flues and chimneys. Wulzen made several views of the rarely photographed neighborhoods beyond downtown in which he showed what conditions were like in the city's residential districts. In one of these views—looking west—his own Upper Market Street district has become the location of a hastily assembled shanty town (pl. 44), and in another Wulzen pointed his camera east, past the City Hall ruin, the Call tower, and the Ferry Building at the foot of Market Street towards the bay and the Oakland hills in the distance to delineate the dramatic scope of the destruction. In this image the return of public transportation on Market Street heralds a resumption of familiar patterns of life (pl. 43).[22]

The precise number of prints Wulzen made of the aftermath phase of the Event is impossible to determine since there currently are no known extant vintage examples of this work. For the same reason, how he made the prints remains open to speculation. He may have printed the negatives in a pictorialist manner similar to his fellow Camera Club members, Stellman and Genthe. During his long membership in the Club, Wulzen did demonstrate interest in the same subject matter as his club associates: he photographed in Chinatown during the years they did; he later turned to the same Yosemite landscape themes they favored; and he photographed the Event as late as the winter of 1906. However, the extant vintage prints of the Chinatown and Yosemite Valley subjects exhibit no pictorialist influences. Wulzen did not mark on or change the negatives in any way, nor do his vintage prints exhibit various pictorialist characteristics and conventions. In the first appraiser's report there is this observation: "D. H. Wulzen's work is all the more remarkable since it is more like the 'straight photography' embraced by a splinter group which much later revolted from the California Camera Club to form . . . Group f/64." And indeed, although he was an active member of the California Camera Club and clearly worked in a pictorialist style, Wulzen's photographic record of post-Event San Francisco marks a clear departure from his earlier, more "artistic" renderings. Instead, Wulzen's prints resemble the documentary-style work of Chadwick, Givens, Worden, and the Pillsbury photographer(s). Like them, he intended his photographs to provide straightforward records of the scenes before him. Because of this, his photographs have the immediacy of news dispatches and stand in opposition to the aestheticizing impulse of the pictoralists' prints. It seems Wulzen was motivated by his own need to see and to know, to make sense of the world through an unmediated directness. The photographs help fulfill a viewer's desire to know the Event, to absorb it, to be taken inside it.[23]

In the many scholarly histories and popular narratives of the Event, writers told the story of an ideal democracy, which, they said, took shape in the first days and weeks. The initial peaceful coexistence among races, ethnic groups, and classes was remarked upon in prose and in verse. Indeed, during the fire and immediately afterwards, there was a willingness on the part of the citizens everywhere to give aid selflessly to fellow sufferers. Official army and navy records confirm reports of looting and fighting among citizens, evidence of the darker side of human nature. But, more often than not, whenever they could, men and women came to the aid of others. As the poet and journalist

Charles K. Field observed, "government bread" and the "same kind of coffee an' crackers" were all rich and poor alike received after hours of "standin' in line." But these democratic conditions persisted for only the first weeks, and by the end of April, pre-Event social conventions were back in place. In Arnold Genthe's view of crowds looking down Sacramento Street at the billowing smoke engulfing the city (pl. 13), a common humanity is bonded by their shared powerlessness and awed contemplation of the fire's relentless surge. Only in a variant photograph of the same Sacramento Street vista does he show viewers that the people gazing at the fire are not just a generalized humanity but also Asian, black, and white individuals as well (pl. 14). But within weeks, when Chadwick and Givens were at work on government-sponsored projects, Asians and blacks had been segregated into separate relief camps; whites were demanding that Chinatown not be rebuilt; black community leaders were openly denied access to neighborhood relief committees; and there was general discontent among the working poor over how easily the city's wealthy elite had come to dominate all relief and reconstruction decisions.[24]

The majority of photographers avoided these harsher aspects of the Event and instead reflected the popular tendency to stress

goodwill and harmony in the community as the efforts to rebuild the city began. In doing so, they helped further the aims of city boosters and obscured a more complete and complex revelation of actual historical events. Willard Worden did photograph two "junk thieves" caught looting a construction site (pl. 48), but dystopic images are rare. In the communal fable about San Francisco that was constructed through words and photographs in the aftermath of the Event, the ideals of a "Phoenix City" were emphasized. In this fable, there was little, if any, criminal malfeasance and no meanness of spirit or social divisiveness. There are no photographs of outraged citizens venting their rage at the prostitutes who were using the Army-built shelters to entertain male "guests" or of the protests outside the temporary St. Francis Hotel where the Mayor's Citizens Committee feted itself and the Red Cross for their relief work. Eventually, the idealized narrative became the accepted history, and even the final tally of the dead was intentionally reduced. Ultimately, the mythologists and the photographers turned the destroyed metropolis into a romanticized terrain for civic instruction and moral introspection, a magnet for amateur image makers as well as professionals.[25]

IV

Since its beginnings as an outpost of American expansion into California, San Francisco had laid claim to mid-nineteenth-century Victorian-era architectural styles and their associated symbolic references. Throughout the city were building types ranging from Periclean Athens to more recent, Roman-inspired Beaux Arts structures. Also on display was an eclectic mixture of Byzantine, Gothic, Islamic, and east Asian-influenced facades—sometimes set down one next to the other within a single city block. In 1906, along with the rest of the United States, San Francisco displayed recently built architectural monuments intended to represent the ideals of western architectural traditions. The popular guide books, most often illustrated with photographs, provided collections of pictures of a uniquely American place defined by its great buildings.[26]

The books and other promotional literature, published by railroad companies, hotels, real estate brokers, and local business groups, derived from the conventions of mid-nineteenth-century photography books of city views pioneered by San Francisco photographers. Three especially noteworthy pre-1906 examples of this genre are *San Francisco: Photo-Gravures From*

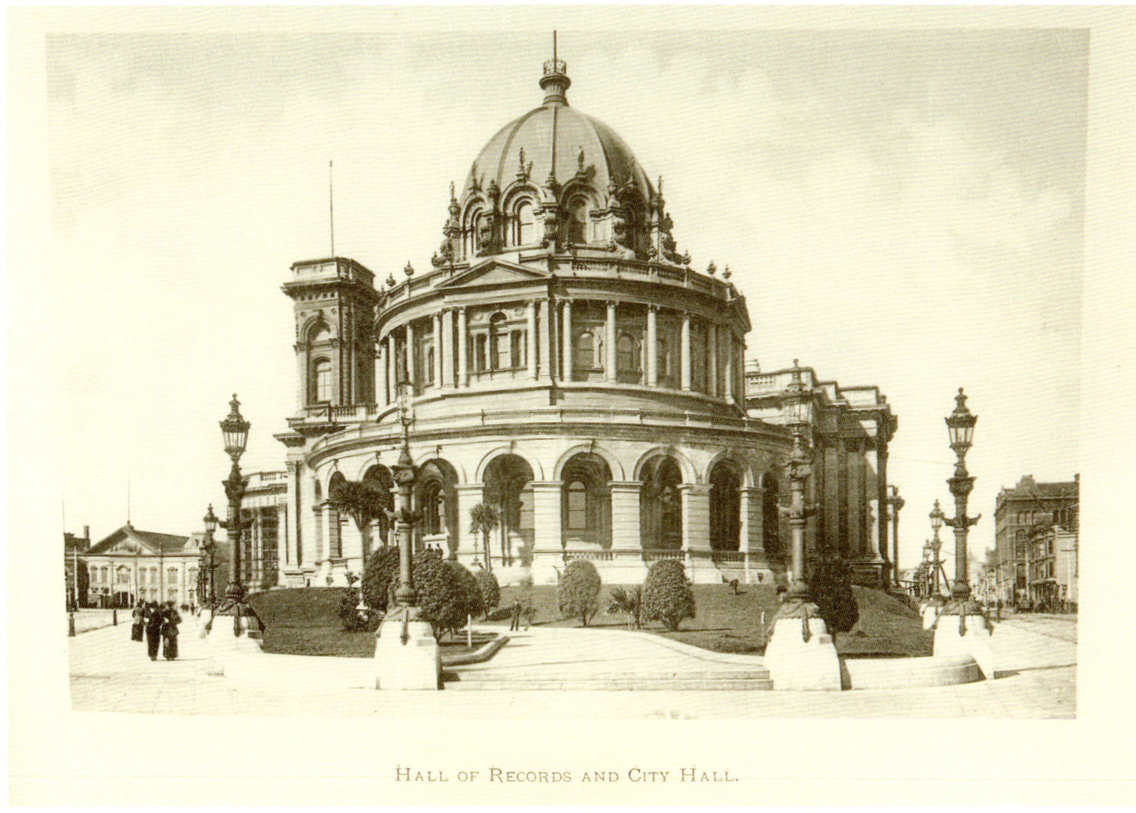

fig. 4. Hall of Records before the building of City Hall (private collection)

Recent Negatives (fig. 4), and two works by Charles Keeler, titled *San Francisco and Thereabout (fig. 5)* and *Souvenir of San Francisco (fig. 6).* In all three publications views of churches, commercial blocks, grand homes, hotels, pleasure gardens and parks, and vistas of major thoroughfares abound. The iconographic character of these books was first established by photographer George Robinson Fardon (1806–1886) in 1856 with his *San Francisco Album,* and continued in the work of Carleton E. Watkins, Eadweard Muybridge, and Isaiah W. Taber. For the photographers working after April 18, 1906, the city's earlier, celebrated persona was inscribed in the same, now destroyed monuments that previously had defined it. It is not surprising then that

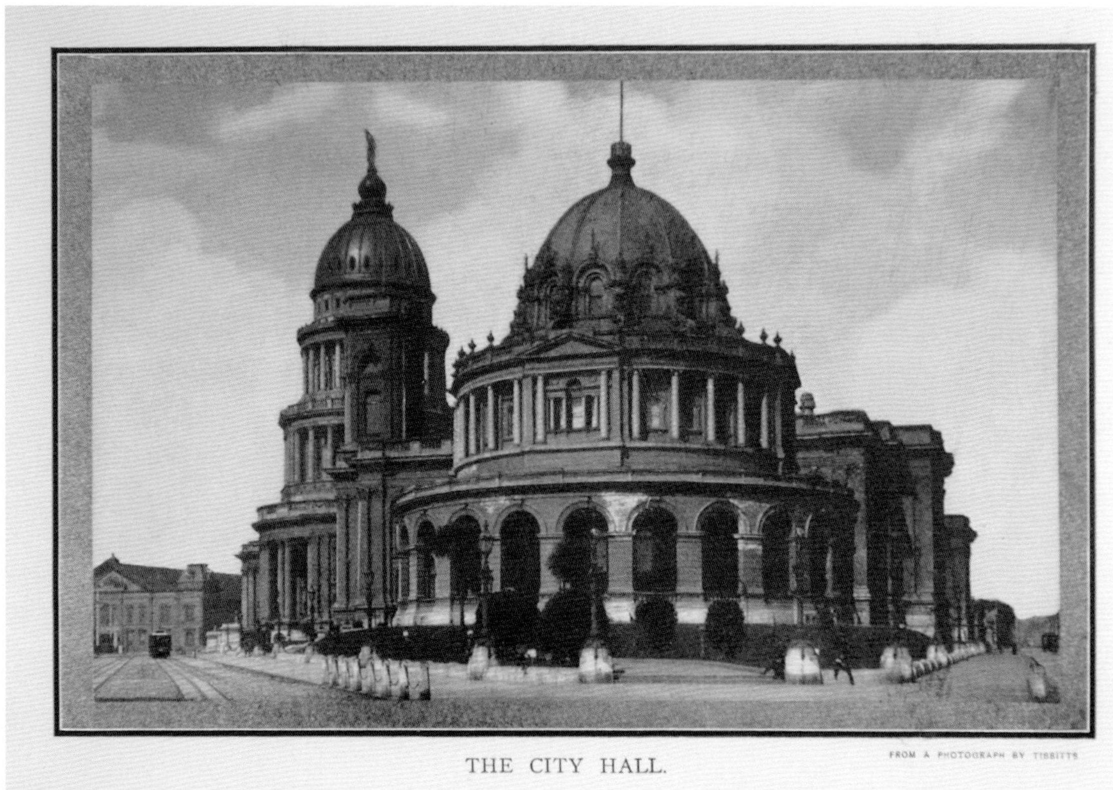

fig. 5. City Hall and Hall of Records on eve of the Event (private collection)

views of identical sites appear in large numbers in the photographs made between 1906 and 1909.[27]

In San Francisco's earlier days, citizens, local boosters, and publishers were eager to proclaim that the depiction of grand architecture provided visual testimony to San Francisco's place in the constellation of great cities. But what had taken over half a century to build lay in ruins after slightly more than a minute of shaking and three days of burning. When the smoke cleared, the scene that presented itself to observers resembled sites some San Franciscans had visited on tours abroad or pored over in photographs and stereo cards. Before their eyes stood scenes reminiscent of the Acropolis, the Roman Forum, and abandoned medieval monasteries—all fashioned in a matter of hours rather than over millennia. For this reason, throughout the remaining spring and summer of 1906, San Francisco became a mecca, the destination for cultural pilgrims who discovered in the ruins the city's melancholy beauty that the earthquake and fire had made visible.[28]

Much like Giovanni Battista Piranesi, whose eighteenth century engravings of Rome's ruins give viewers an encyclopedic visual record of the ancient capital, San Francisco photographers examined the bleak ruinscape, confronting it in exacting detail. After the initial popular desire for the "smoke pictures" subsided, there emerged a demand for photographs of the destroyed structures. San Francisco's buildings, in their original state, had glorified the city in which they had been erected, and they brought praise to the architects who designed them and the investors who provided the financing. The ruined buildings were ministers of different emotions, however. Precipitate of temblor and fire, they were physical proof that unforeseen natural forces could turn an entire city into acres of rubble in a matter of hours. As with Rome, a broken San Francisco too could instruct tourist and viewer. Considerations of human fallibility, the uncertain fates of entire societies, and, for some, the very meaning of life itself seemed to be written in the hieroglyphs of fallen brick and twisted steel. The destroyed structures fulfilled the popular yearning for encounters with picturesque landscapes and their peculiar instructive value. The ruins offered a lesson for people who knew the histories of antiquity, the middle ages, and the more recent nation-states of post-feudal Europe. History pointed to a humbling moral: great states and their citizens wax for a while, but inevitably wane. The wanderers in Piranesi's views contemplate the transitory nature of all things, how in a civilization's greatest achievement lies concealed its demise. The San Francisco ruins were similar messengers for the contemporary audience.[29]

Commenting on the "classical scale" of the ruins in Rome, Piranesi's ruins, Christopher Woodward, a scholar of ruins and their histories, observes how "the more magnificent the edifice the more effectively its skeleton demonstrated the futility of mortal pride." The San Francisco ruins were also magnificent in scale, and they spread over hundreds of acres. The photographing of San Francisco ruins also was extensive, and thousands of prints resulted from the numerous photographers' work. The most prominent subject in this large body of work is the City Hall, the dominant physical artifact and visual presence in the cityscape. Since it would not be demolished until March 1909, almost three years after the Event, City Hall remained the greatest palpable evidence of an earthquake's power

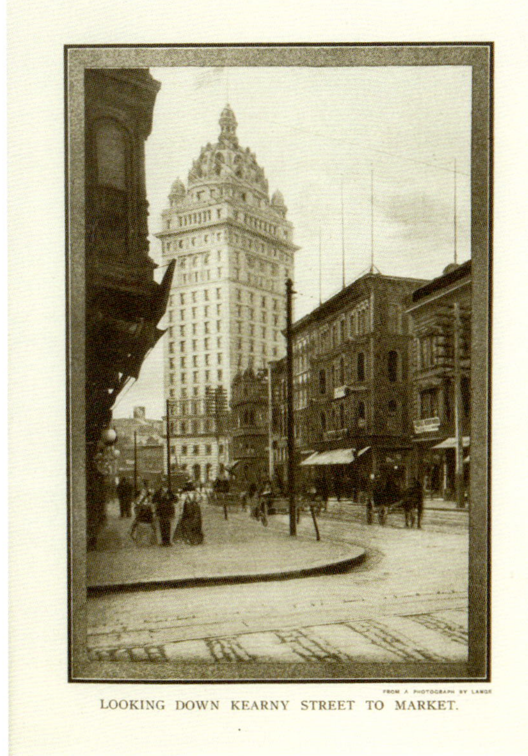

fig. 6. *The Call Building before earthquake and fire (private collection)*

to rip apart buildings and rupture the normal course of events. Even when the subject was a panoramic view of the blighted cityscape, as in Genthe's "Steps that Lead to Nowhere" (pl. 26), or Worden's "Portals of the Past" (pl. 27), City Hall is the structural and compositional feature commanding the viewer's attention.[30]

When Mayor Schmitz convened the first meeting of the Citizens Committee, he included representatives of the city's large corporations and real estate interests. At the Committee's first meeting on the second day of the fire, discussions regarding a rebuilding plan were first voiced. Only seven months before, in September 1905, a gala banquet had been held to praise the grand *Plan for San Francisco,* which Chicago architect and planner Daniel H. Burnham had presented to the audience. Burnham's vision combined a rationalized street plan on the Parisian model with grand neoclassical governmental, cultural, and architectural complexes. Most of the men there that night eventually served on the Citizens Committee or one of its various permutations over the following months. The most powerful subcommittee, the Finance Committee, was created in May and ultimately decided the direction rebuilding would take. It was dominated by one-time Burnham plan supporters who were now its foes. The Finance Committee chairman, James D. Phelan, had previously been an enthusiastic Burnham plan supporter and had contributed a "Historical Sketch of San Francisco" to Burnham's official report to the city. But despite his long friendship with Burnham and a commitment to "city beautiful" planning, Phelan ultimately supported the rejection of Burnham's grand design for San Francisco. Instead, Phelan and others supported rebuilding efforts that would focus on the return to "normal" business—that is, a return to pre-Event patterns of shopping and commerce—as quickly as possible. Burnham came to San Francisco in early May to encourage local supporters of his plan. He admonished anyone who would rush to recreate the city that was, and instead advised a steady implementation of the plan they had recently embraced just months before.[31]

In 1905, when Burnham presented the drawings which Edward Bennett, his draughtsman, had prepared, he estimated that it would require decades of incremental work before the complete plan would be realized. He acknowledged that metropolitan rhythms of life should be disturbed as little as possible. However, now that the earthquake and fire had razed hundreds of acres of the city, what would have required five decades could be accomplished in half that time. Burnham reasoned that the replacing of infrastructure and building programs could be undertaken simultaneously. And, since almost half the city's population was living in refugee camps, implementing his plan would make their lives better, not disrupt them. But neither the populace nor the city leadership agreed, and the opponents of the Burnham plan were victorious. The plan's supporters did gain one significant concession: the grand civic center in Burnham's original plan would be built. Not in the

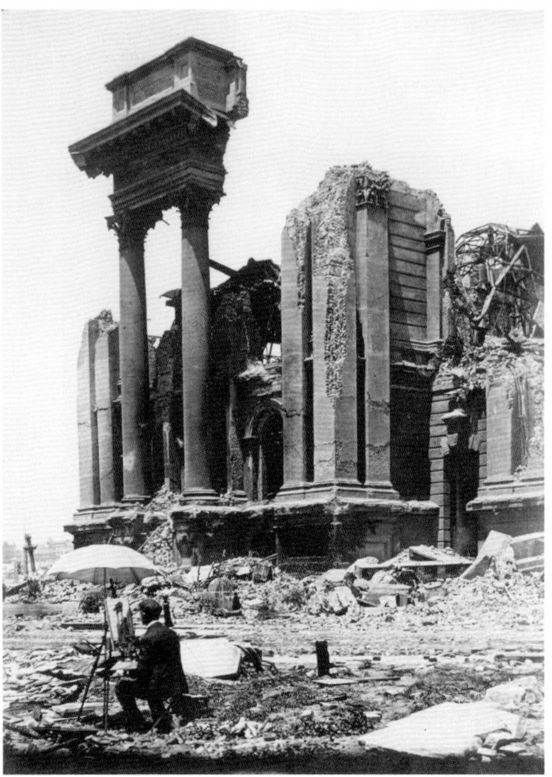

fig. 7. Painter at work on the City Hall ruins
(Huntington Library)

vicinity of destroyed City Hall, however, but two blocks west and a block north of the originally identified site. Although this quick decision decided the fate of the pre-earthquake City Hall site, the ruined building would linger there for nearly three years. It was not until October 1909 that the last remnants of the Mayor's original Citizens Committee hosted the Portola

Festival, San Francisco's official celebration of closure and renewal. In early March, six months before Portola and almost three years after the temblor, the dismantling of City Hall began with the lowering of the monumental female statue that had stood on top of its dome for a decade. A crowd gathered below to watch the twenty-two-foot-tall bronze figure descend to earth. The wind was vigorous and tragedy struck. The steel rope corset wrapped around the goddess began to twist and gouge into the bronze. There was a terrible groaning sound as the trunk was severed from the appendages. Louis J. Stellman made a poignant photograph of the brutalized, once proud symbol of metropolitan spirit (pl. 51).[32]

All the photographers of the Event recorded the widespread devastation in San Francisco, but only a small minority carefully examined and photographed particular ruins. That is, the great scope and grandeur of destruction were the primary subjects of most, but the revealing details left behind by the temblor and fiery holocaust received the consideration of only a few photographers: Edgar A. Cohen, one of the first to take note of the relationship between the scenes he observed in San Francisco and the ruins of earlier, long past eras; Arnold Genthe, who had earned a doctorate in Roman and Greek Studies before he turned to photography; Willard Worden, a sensitive artist throughout his long career in San Francisco; and Louis J. Stellman, author of *The Vanished Ruin Era*. This group best represents the photographers who thoughtfully investigated the ruins of San Francisco. In early summer 1906, in *Camera Craft,* Cohen wrote that most photographers included "more than the desirable part[s] on the plate" when they made their prints, and thereby "destroy[ed] the feeling that [the view] might be a scene in ancient Greece." Cohen was already interested in the aesthetic possibilities of the destroyed cityscape. Soon, painters and print makers also began to extract particularized details from the panoramic view *(fig. 7).* Artist Joseph Penell traveled from New York to add San Francisco to his "Cities Series" after rebuilding had gotten underway, but while "there still remain[ed] some reminders of the epochal disaster" *(fig. 8).*[33]

Stellman, who joined the California Camera Club in 1902, was a friend of Wulzen, Genthe, and other members. He remained active in club affairs for the next decade and a half, and it comes as no surprise that his *The Vanished Ruin Era* reflected pictorialist photographic beliefs and modes when it was published in 1910 *(fig. 9,* pls. 16 and 49–51). Stellman's response to the ruinscape was as profound as that of any of the visual artists. He began photographing the Event while the fire still raged and the evacuation of

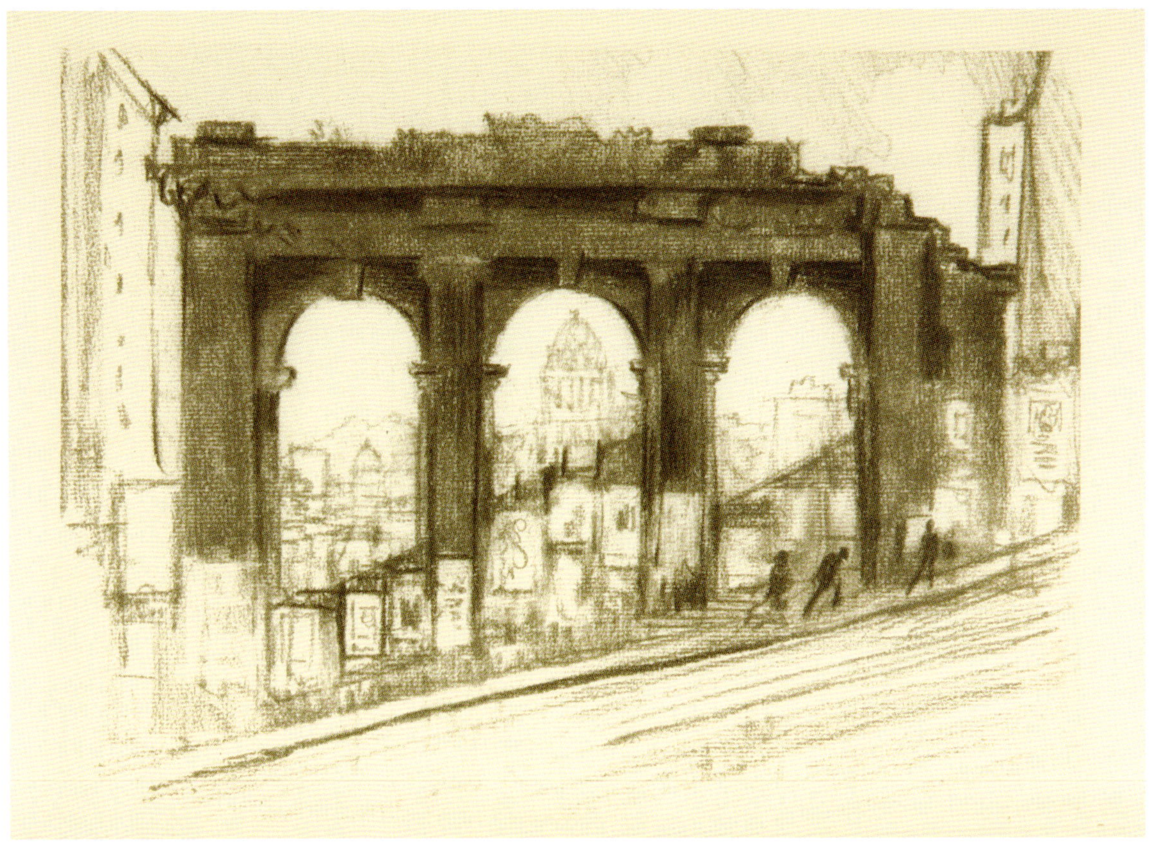

fig. 8. Joseph Penell's fanciful vista of San Francisco: Twin Peaks, Humboldt Bank, and City Hall (private collection)

the city's inhabitants was taking place (pl. 16). He continued to observe and explore the city over the next three years, examining life in the camps (pl. 50) and, of course, the demolition of the City Hall and the resulting demise of the "Goddess of Progress" (pl. 51), about which he wrote: "Ah, gruesome jest of fate! that I have foiled/God's mighty elements, to end my span/ Of life—a vandal's prey—to be despoiled/of being by the hand of puny man!" *The Vanished Ruin Era* is a compilation of his three-year photography project with the negatives reprinted as photogravures carefully prepared for publication in pictorialist style.[34]

Over the course of the three-year long Event, Stellman worked at several San Francisco newspapers: the *News,* the *Globe,* and the *Post.* His assignments demanded a daily encounter with the unadorned facts of life after April 18, and he reported on the destruction, the rehabilitation, and the reconstruction in detail. His book, however, was a post-Event phenomenon, a recapitulation. By the time *The Vanished Ruin Era* was published, the Portola Festival had come and gone. The celebration marked the symbolic end of the Event and the beginning of a new era in the city's history. In his reconsideration of the photographs he had made in the years and months leading to Portola, Stellman understood that just as there had been a somber beauty in the ruins, so too was there a tragic comeliness surrounding the entirety of all that had happened between April 1906 and October 1909. To turn the scenes he had photographed into artfully made photogravures and then to weave poems among them was his declaration of closure and an affirmation that the gravures and poems could bring forth a truth, as had his newspaper reports. Images, words, and memories evoked the past, but they could not recall it. The Event was now concluded. But Stellman's artistic presentation ensured its endurance in time. The circle closed. A cycle had been endured. Newspaper journalist, photographer, and poet, Stellman had seen

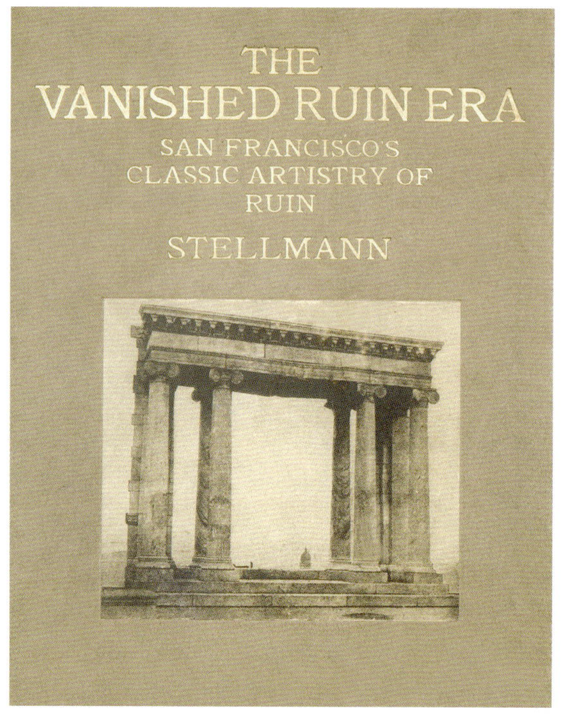

fig. 9. Cover of Lewis Stellman's photographic and poetic paean to San Francisco (private collection)

things fall apart and then had watched them be put back together again. The ruins had vanished before his eyes. *The Vanished Ruin Era* memorialized the Event and how San Franciscans had lived through it. But now they were safe, no longer dispossessed.

The photographs of the shaken, burned, and recovering city chronicle the enormity of the loss: lives lost; homes lost; families destroyed; familiar landmarks gone; and, for a time, the city

itself seemed lost. The very fabric of life, the lines and threads humans follow as they travel from one place to another and back had been stripped away. Photographs of the rebuilding phase of the Event focus on the insistence upon and measured success of starting over. Fire and "smoke" photographs and the photographs of toppled structures were not only about loss. They spoke of dislocation as well. San Franciscans believed—often were cajoled into believing—that their real guarantee of safety lay in reconstructing the city as it had been. Although reason showed this was impossible, that they could in fact not go home again, they desperately wanted to. The effort to reconstruct the original map, and the city it recreated, was Herculean. The human energy invested clearly was tremendous.

"Resurgam! Let the new San Francisco burst forth from the ruined shell of what had been," wrote Pierre Beringer in an article for the *Overland Monthly* a year after the earthquake. Beringer was astounded at the rapid pace of reconstruction already underway. But just as individual victims of trauma struggle to put shattered lives and bodies together, so too do social entities. Individuals reconstructed their lives, and the community also began to heal. Slowly, new landmarks took the place of those that had disappeared, new details of infrastructure replaced the old ones now gone and missing, and eventually a new map emerged to help the disoriented on their way. While humans grieve, they also plot to rebuild.[35]

The resurgence of San Francisco and the surrounding temblor-struck areas is a testimony to the spirit of the communities and to the individual belief in an American dream of progress and national prowess. The rise of a "new" San Francisco out of the debris and ashes also underscores the power of capital and the spirit of capitalism, in the pre–World War I California. The dream of Burnham, often a partner of prominent capitalists himself, had been undone by local business interests and eager investors from the United States and abroad. Back to business as it had been—and where it had been—was a prescription for recovery people read and were told. They acquiesced. And San Francisco after the disaster of 1906 is, a century later, surprisingly like the city of April 17, the day before the temblor struck and fire began.

As the rebuilding proceeded, photographers continued to tell the story of the Event through images. In particular, four photography projects represent the medium's relationship to the culture and the historical moment in the last years of the Event: The R. J. Waters Company photographic documentation of the Phelan Building construction; an album assembled by a local dentist and amateur photographer, Crittenden Van Wyck; the *Portola Souvenir,* a city-authorized publication presenting views of reconstructed city buildings; and Willard Worden's sequence of photographs of the dismantling of the old City Hall.[36]

V

Construction on the Phelan Building, one of the earliest symbols of a reinvigorated San Francisco, began in late summer 1907. On October 7, after the foundation work had been completed, an R. J. Waters Company photographer began the documentation of the building process at two-week intervals, ending with the photograph dated September 19, 1908 (pls. 52–55). As the structure took shape, the sequence of photographs shows that its architectural heritage is "Chicago Style" both in engineering and design. Constructed with a light-weight steel skeleton and wrapped in a "Renaissance/Baroque"-patterned skin of baked terra cotta tiles, the Phelan Building was a successful articulation of the tall office buildings that had begun to define the city's skyline in the late 1880s. Set on the triangular plot formed by the intersection of three streets, Market, O'Farrell, and Grant, it was a West Coast reincarnation of New York's Flatiron Building. Daniel H. Burnham, designer of the New York building, had guided it to completion in May of 1907, a scant three months before foundation work began for the San Francisco structure. The Flatiron Building too has a "Renaissance/Baroque"-patterned skin, although in this instance the material is granite. Also, like its San Francisco counterpart, the Flatiron Building occupies a triangular lot created by the intersection of Broadway, Fifth Avenue, and Twenty-third Street. Burnham and his partner John Wellborn Root (1850–1891) were the architects who brought tall office buildings to San Francisco (the Chronicle Building in 1889 and the Mills Building in 1891 [pls. 23 and 40]). Burnham had no direct relationship with the Phelan Building, and at the time of its construction his involvement in San Francisco affairs was minimal. Burnham's imprint, nevertheless, is on the Phelan Building project. The building's owner, James D. Phelan, chair of Finance on the mayor's Citizens Committee, was the scion of a wealthy San Francisco family and a prominent member in the city's conservative business community. He had served as a member of California's delegation at the world's fair in Chicago in 1893 and had been instrumental in bringing some of that fair's exhibits to San Francisco's Golden Gate Park for the California Midwinter International Exposition of 1894. In this setting his friendship with Burnham began to flourish and Phelan encouraged the local City Beautiful organization to invite Burnham to develop his San Francisco Plan. Phelan admired Burnham and eagerly embraced the building of a structure closely modeled on the latter's Flatiron Building.[37]

The R. J. Waters Company photographs mark the Phelan Building's physical aspect on twenty-four selected days of its young life, and, like a parent's marks on a wall record a child's growth. The photographs chart the structure's beginning below the city's surface, the burst of growth during the structure's first summer, and its solid maturity—ready for business—in September 1908. It took just over a year to construct the Phelan Building, and its completion signaled that a level of more normal activity had returned to Market Street, the city's main thoroughfare, the street San Franciscans called "the main stem." From Twin Peaks to the Bay, "San Francisco Resurgam" was becoming more reality than prophecy. The great eleven-story building was a visual marker that illustrated the new urban façade and infrastructure. Its ornate skin linked it to the pre-Event city, while its steel skeleton was a harbinger of modern San Francisco. And in fact, modern infrastructures were now being constructed throughout the city, replacing the now destroyed post–Civil War technologies. As a result, for some time San Francisco could claim to be America's most modern city. However, architectural styles continued to resemble the more familiar face of Edwardian times, the inheritance of the era labeled "Victorian." Even painter and arts and crafts designer Arthur Matthews, champion of new ideas in the arts, could find comfort in the Phelan Building and the partially rebuilt city it graced. In his journal of ideas, *Philopolis,* he wrote: "[T]hat San Francisco has not reformed itself on strictly utilitarian lines, has not forgotten a certain comeliness, as an essential to every work, is proof that the inhabitants still retain [an] aspiration to be more than utilitarian."[38]

The "comely" San Francisco, which the Phelan Building helped bring about, was taking shape by the end of 1908, and only a few physical reminders of the wreck and ruin wrought by temblor and fire remained. It was in an atmosphere of civic pride and celebration engendered by rebuilding that Crittenden Van Wyck, a dentist and snapshot enthusiast, undertook his photographic record of the city in which he was born and had lived for most of his adult life. He began to wander through the city and take pictures in 1907. Van Wyck's modest prints (all are 3" x 4") indicate that he had not fully mastered the medium. The images are frequently poorly focused and the prints are often marred by in-camera or darkroom errors. Despite their shortcomings, the Van Wyck photographs are compelling artifacts from the second year of the Event. Van Wyck made photographs of locations and structures previously established as visual staples in the vernacular tradition of urban images—the Ferry Building, the abandoned shell of City Hall, Union Square Park, the downtown Financial District, and parts of the erased Victorian San Francisco where modern structures were now rising. He also marked the locations, both figuratively (pl. 56) and literally (pl. 57), where his personal history intersected the Event's communal narrative.[39]

In 1941, three decades after he had undertaken his project, Crittenden Van Wyck assembled an album of selected images. In a brief letter he attached to the album Van Wyck shared biographical data (born in San Francisco in 1870, U. S. Army dentist from 1917 to 1919) and noted his decision to give "[one hundred and ten] Kodak pictures" to the San Francisco Public Library in October. He began editing the album in May 1941, a month after the thirty-fifth anniversary of day one of the Event. (The survivors of the disaster commemorated the Event annually, but after World War I the five-year milestones especially were celebrated.) Van Wyck was seventy-one years old when he donated the album to the library, one of the buildings shaping the Civic Center remnant of the Burnham plan. The occasion almost certainly provoked the consideration of his own mortality as he assembled the album. The album's earliest photographs date

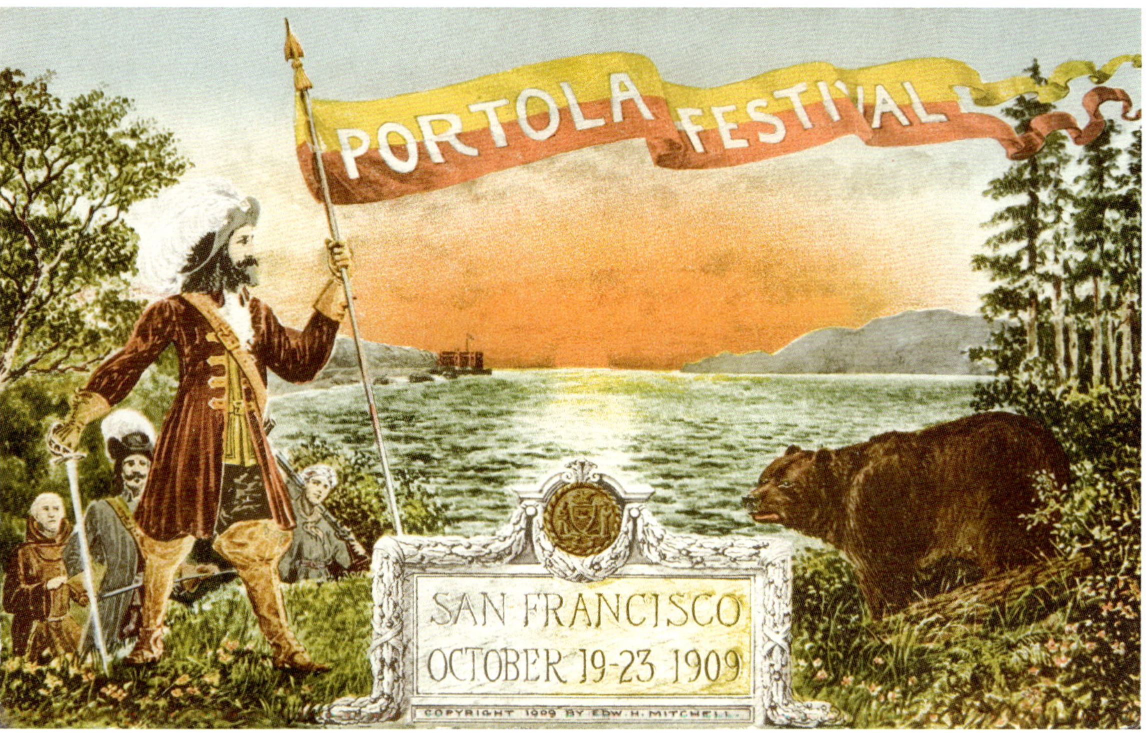

fig. 10. Portola Festival souvenir postcard (private collection)

beams and piers, and the numerous photographs of various buildings under construction analogize the Waters Company Phelan Building sequence made in the same year. Confidence in the rising city, the photographs say, has displaced the widespread grief and despair of the previous years. The last traces of that more sorrowful era were carted away as the city prepared to mark the three-year anniversary in April 1909 and as the old City Hall began to be taken apart (pl. 51).[40]

The Portola Festival (October 19–23, 1909) was intended to be the expression of an end to political feuding and the beginning of a new, rebuilt metropolis. P(atrick) H(enry) McCarthy, the successful compromise candidate for mayor in the November election, pledged to bring to San Francisco an administration "tolerant" of opposing political viewpoints and eager to establish a citywide "get-together spirit." The festival was the first indication of the social amity and proof, McCarthy said, that the city had become the "Paris of America."[41] The Portola Festival was a success, and a committee to bring a world's fair to San Francisco held its first meeting two weeks before McCarthy won the election. Frank M. Todd, historian of the Panama Pacific International Exposition of 1915, called the Portola Festival a celebration of as great a "victory of man over loss and discouragement and vast disaster as had [ever] been exhibited to

from early in 1907 and culminate in late 1908, two event-filled years for Van Wyck and his fellow survivors. In 1907 the city had suffered an outbreak of bubonic plague, a bloody streetcar strike, and an eruption of political acrimony around the graft trials. Then, in 1908, Fremont Older, publisher of the *San Francisco Bulletin,* was kidnapped, and the graft trial prosecutor, Francis Heney, was shot in open court. The political strife became unbearable and culminated in a forced reconciliation of the contesting factions.

Meanwhile, the physical regeneration of the city was palpable, and this proof of recovery became the rallying point for pleas of social renewal too. Van Wyck's modest album visually echoes the claims of the city's resurgent state. His photographs, despite the flawed technique, mirror the work of the other photographers. He shared their interest in the assembling of steel the world before," and he noted that "the Portola Festival had been vital to the development of the 1915 World's Fair." The citywide carnival included a grand parade, athletic competitions, a beauty pageant, an automobile race, international fleet displays in the bay, and Mardi Gras–like pageantry *(fig. 10).* The defining theme of the five-daylong festival was made evident in its name: Gaspar DePortola had led the first Spanish exploration party to the Bay Area in 1769 *(fig. 11).* Associating Portola with the 1909 celebration tied the city's founding one hundred forty years earlier with its rebirth after near obliteration in 1906. The Portola Festival marked the end of one era and the beginning of another.[42]

On October 21, 1909, the Department of Public Works of the City of San Francisco issued its *Portola Souvenir,* which represented the city government's participation in recreating the built environment and was meant to "call attention to the great work undertaken and now in progress" *(fig. 12).* The *Portola Souvenir* consisted of "sixty-four pages of halftones [sixty-seven images] showing school houses, fire department buildings, and other structures completed or in various stages of erection." Also included are photographs of "cisterns, fire boats, sewers, street pavements . . . and other classes of public work" (pls. 58–65). The *Portola Souvenir* is

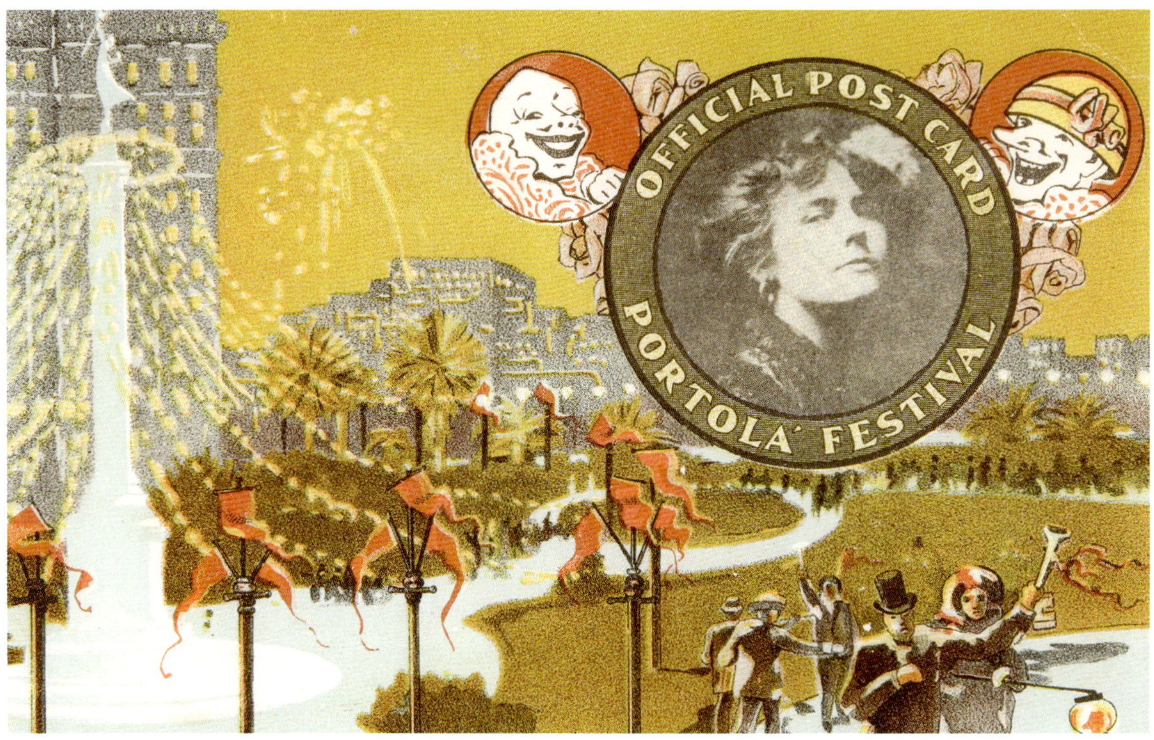

fig. 11. Portola Festival souvenir postcard (private collection)

the visual heir of Fardon's *San Francisco Album* published fifty years before the Event. Fardon's book represented the efforts of mid-nineteenth century city boosters to advertise San Francisco's social well-being and physical vigor despite the fact that the city was in the grips of a violent civil insurrection. When Fardon and his publishers assembled the *Album,* they included photographs of volunteer firehouses, school buildings, and significant public works. They were eager to assure their audience that the city had transcended its recent political turmoil and was a safe haven for investment dollars and immigrants. The *Souvenir* served a similar purpose: it would call attention to the great work undertaken and now in progress after the great earthquake. The *Souvenir* was intended to assuage any economic and political uncertainties

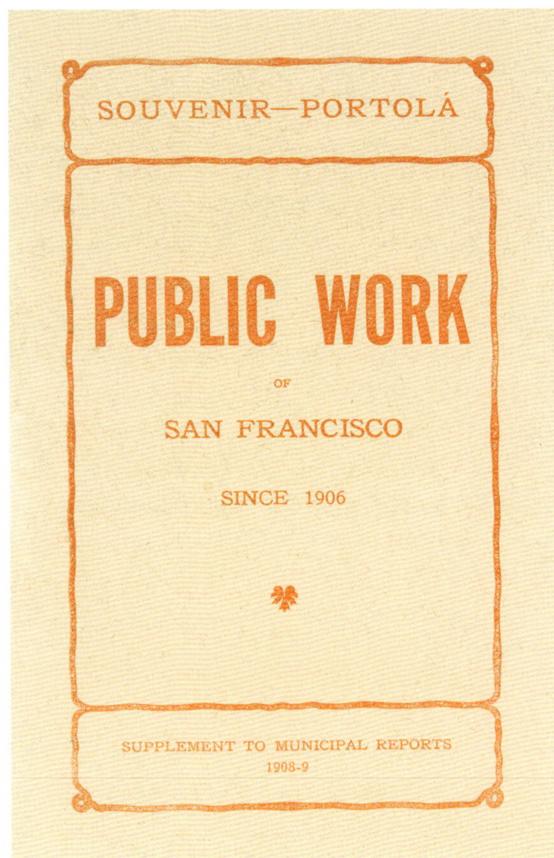

fig. 12. Cover of city-published pamphlet celebrating a rebuilt San Francisco (San Francisco History Center, San Francisco Public Library)

regarding the city's rehabilitation after a lengthy three-year effort.[43]

On April 21, 1906—the third day of the fire—Will Irwin, a San Francisco native transplanted to New York, wrote an "obituary" for his beloved city. Originally published in the *New York Sun,* the newspaper for which he was working, Irwin later recast the article as an independent essay that he titled *The City That Was.* It is a tribute to the city he remembered as a youth and which was now, according to the reports arriving in the *Sun's* editorial room, almost completely destroyed. "The old San Francisco is dead," began Irwin, "[t]he gayest, lightest hearted, most pleasure loving city of the western continent . . . is a horde of refugees living among ruins." The remainder of the essay is Irwin's description of the city's neighborhoods, its landmarks, idiosyncratic characters, geography, and history. It is a composite portrait in words, colored by Irwin's romantic glance back at the city that had "mingled the wine of her bounding life with the wine of his youth." No photographs accompanied the published piece, but Irwin's descriptions of San Francisco remarkably resemble the visual descriptions of the city found in the photographs Arnold Genthe had published in the 1890s in the magazine, the San Francisco *Wave.* At the time of publication, Irwin was serving as the magazine's editor, and Genthe's images were not only seen by Irwin, but quite possibly approved by him for publication.

In both Genthe's photographs and Irwin's essay the implied relationship between word and image becomes explicit. Both are descriptions, and when they appear together, the reader's knowledge of what is being described is doubly enriched. Proverbially, a picture is worth a thousand words, but to ensure the proverb's accuracy—and the image's value—we need those words before we can determine that picture's value as part of a usable past. The marriage of words and photographs provides us with something new, a precipitate the two entities standing alone do not provide. In the descriptions he has written here for each of the following photographs Marvin Nathan demonstrates how important and valuable is this linking of photograph and word. Both methods of description are illustrative and together become illuminations. In his commentary he has subjected the photographs to the kind of close reading they deserve and, in most instances, never have received. In conjunction with the photographs he has demonstrated how, when historical and architectural details are connected to a visual text, the historian's "redemption" of the past begins to reveal itself.

Rodger C. Birt
Oakland, California

NOTES

1. On the relationship of memory with photography see Roland Barthes, *Camera Lucida: Reflections on Photography* (New York: Hill and Wang, 1981) and Russel B. Nye, "Notes on Photography and American Culture, 1839–1890" in Louis J. Budd, ed., *Toward a New American Literary History* (Durham: Duke University Press, 1980); Siegfried Kracauer, *History: The Last Things Before the Last* (Oxford: Oxford University Press, 1969), 31, and *Theory of Film: The Redemption of Physical Reality* (Princeton: Princeton University Press, 1960), 10. See also Dagmar Barnouw, *Critical Realism: History, Photography, and the Work of Siegfried Kracauer* (Baltimore: Johns Hopkins University Press, 1994).

2. For a discussion of photographers other than those I include here see Rodger C. Birt, "Envisioning the City: Photography in the History of San Francisco, 1850–1906," (PhD diss., Yale University), 1985, 349–369; most prominent among the amateurs were novelist Jack London and his wife, Charmian, who shared their experiences with the public in the May 5, 1906, issue of *Collier's* magazine.

3. Carl-Henry Geschwind, *California Earthquakes: Science, Risk and the Politics of Hazard Mitigation* (Baltimore: Johns Hopkins University Press, 2001), 20. Geschwind says the 1906 event was "the most expensive natural disaster to strike the United States until Hurricane Andrew ripped through Florida in 1992."

4. *Daily Humboldt Standard*, April 18; William Bronson, *The Earth Shook, The Sky Burned* (Garden City, New York: Doubleday & Co., 1959), 31.

5. About time: the time the 1906 temblor was first reported at different locations, as noted in this essay, is based on later newspaper reports. F(usakichi) Omori, the Japanese geologist, came to California in May and spent the rest of that year investigating the earthquake and its effects. He writes, "(T)he times of earthquake occurrence observed at the California University and the Lick Observatory were respectively 5*h* 12*m* 39*s* and 5*h* 12*m* 12*s* A.M. (Western States Time, or that of longitude 120° W.); the time of commencement of the disturbance at the origin itself being probably about 5*h* 12*m* A.M." See F. Omori, "Preliminary Note on the cause of the California Earthquake of 1906," in David Starr Jordan, ed., *The California Earthquake of 1906* (San Francisco: Robertson, 1907).

6. William James, "On Some Mental Effects of the Earthquake," *Menninger Perspective* 1, no. 1 (November, 1990); Bronson, *The Earth Shook*, 132.

7. Bronson, *The Earth Shook*, 140.

8. *Fresno Morning Republican*, April 19; *Gilroy Advocate*, April 21; *San Jose Mercury*, April 20.

9. Tremors were recorded in Tokyo and Germany with highly sensitive measuring devices. Beyond Hollister, near the mission, residents there said the movement was "slight," and there were no "authentic reports of any shaking," Charles Derlerth, Jr., "The Destructive Extent of the California Earthquake of 1906," in Jordan, *The California Earthquake of 1906*, 246.

10. Edgar A. Cohen, "With a Camera in San Francisco," *Camera Craft* 12, no. 5 (June 1906), 185.

11. Arthur C. Pillsbury was a photographer and a clever entrepreneur who purchased the rights to publish other photographers' work under his own imprint; *New San Francisco Magazine*, no. 1 (May 1906); Melinda Pillsbury-Foster, "A Glorious Adventure in Images: The Life of Arthur C. Pillsbury," *The Argonaut* 18, no. 1, (Spring, 2007), 52–65; Charles Keeler, *San Francisco Through Earthquake and Fire* (San Francisco: Paul Elder and Company, 1906), 19. Keeler's account is one of the best of the first group of eyewitness narratives.

12. Arnold Genthe, *As I Remember* (New York: Reynal and Hitchcock, 1936), 89;

Toby Quitslund, "Arnold Genthe: A Pictorial Photographer in San Francisco, 1895–1911," PhD diss., The George Washington University, 1988, 249–283; much remains to be unearthed regarding Willard E. Worden (including his relationship to Clinton E. Worden—see plate 27 and fig. 9); also see Michael Bowen, "The Rediscovery of Willard Worden," *San Francisco Sunday Examiner-Chronicle [California Living* section], Feb. 22, 1981, 34–36; "Photographs of San Francisco by Willard E. Worden," *Popular Photography*, (August, 1977), 138; see also notes and related items regarding Worden in the Wells Fargo Bank History Dept. Willard E. Worden collection.

13. John P. Young, *San Francisco, A History of the Pacific Coast Metropolis* (San Francisco: S. J. Clarke, 1912), 866.

14. There were twenty-two aftershocks by noon, *Napa Register* May 4, 1906, and eight more before 7:00 P.M., Lawrence Kinnard, *History of the Greater San Francisco Bay Region* (New York: Lewis Historical Publishing, 1966), 2, 122. Kinnard says tremors were recorded at the rate of two a day into early May; Genthe, *As I Remember*, 88.

15. Peter Palmquist, *Carleton E. Watkins: Photographer of the American West* (Albuquerque: University of New Mexico Press, 1983), 83.

16. For the official United States Army version of the dynamiting in San Francisco see "Report of Brig. Gen. Frederick E. Funston," in Maj. Gen. Adolphus W. Greely, *Earthquake in California April 18, 1906: Special Report* (Washington D. C.: Government Printing Office, 1906), 5–11; Albert S. Reed, "The Conflagration in San Francisco; Special Report to the National Board of Underwriters," (New York: 1906), 11 and Mary Austin, "The Temblor, a Personal Narrative," *Out West* 24, no. 6 (June 1906), 506, both report a heavy rain falling on the night of the twentieth. A more critical analysis of army dynamiting activity is developed in Grove Karl Gilbert et.al., *The San Francisco Earthquake and Fire of April 18, 1906, and Their Effects on Structures and Structural Materials*, U.S. Geological Survey Bulletin no. 324 (Washington: Government Printing Office, 1907).

17. Funston temporarily was in command of the Pacific Division, U.S. Army from April 18 to April 22; Greely reassumed command on April 23; Col. Charles Morris was commanding officer of the Presidio; Badger was commander in chief of the U.S. Navy Pacific Squadron.

18. There are sixty-six vintage Chadwick prints in the collection of the Visual Studies Workshop, Rochester, New York: the analysis of Chadwick for this study was based on the workshop collection. Although notes in the workshop collection identify him as "Harry W.," according to his obituary his legal name was Henry W. Chadwick; Vallejo, California *Times Herald*, Jan. 31, 1933; four additional vintage prints are in the collection of the History Dept., Oakland, Calif. Public Library—Main Branch; for James D. Givens see James D. Givens, *San Francisco in Ruins: A Pictorial History of Eight Score Photo-Views of the Earthquake Effects Flames' Havoc Ruins Everywhere Relief Camps.* (San Francisco: L.C. Osteyee, 1906). Wulzen's career as a pictorialist photographer in the California Camera Club received an in-depth analysis in Anthony W. Lee, *Picturing Chinatown: Art and Orientalism in San Francisco* (Berkeley: University of California Press, 2001), 136–143; C. Cameron Macauley, *An Independent Appraisal Report of a Collection of Vintage Glass Plate Photographic Negatives*. Macauley made the appraisal for the San Francisco Public Library History Dept., which has the thirty-five Wulzen glass plate negatives depicting the earthquake and its aftermath I examined for this study; for discussions on aspects of Stellman's career see Richard Dillon, *Images of Chinatown: Louis J. Stellman's Chinatown Photographs* (San Francisco: Book Club of California, 1976) and Lee, *Picturing Chinatown*, 181–199.

19. Young, *San Francisco*, 836; for a history of the Citizens Committee, its permutations, and the various subcommittees, see Judd Kahn, *Imperial San Francisco: Politics and Planning in an American City, 1897–1906* (Lincoln: University of Nebraska Press, 1979), 128–153.

20. Greely, *Earthquake in California*, 37.

21. Givens, *San Francisco in Ruins*; Michael G. Wilson and Dennis Reed, *Pictorialism in California: Photographs 1900–1940* (exh. cat.), J. Paul Getty Museum, Sept. 13–Nov. 27, 1994, 1–22.

22. Lee, *Picturing Chinatown*, 136–137; On pictorial photography during these years see Wilson and Reed, *Pictorialism in California*, 1–22.

23. Macauley, *An Independent Appraisal*, 6.

24. Charles K. Field, "Barriers Burned," in Arthur Chandler, ed., *Old Tales of San Francisco* (Dubuque, Iowa: Kendall/Hunt, 1987), 282–283; there were two camps for Chinese refugees, one in Oakland and a second on the Presidio grounds, Greely, *Earthquake in California* 46; black-operated newspapers in the east devoted significant coverage to the fate of the Bay Area's nonwhite community, see *New York Age*, April 26, 1906 and May 3, 1906; *Baltimore Afro-American*

Ledger, April 21 and 28, May 5, 12, and 26, 1906; Bronson, *The Earth Shook*, 127; William Issell and Robert W. Cherny, *San Francisco, 1865–1932: Politics, Power, and Urban Development* (Berkeley: University of California Press, 1986), 155–158; Michael Kazin, *Barons of Labor: The San Francisco Building Trades and Union Power in the Progressive Era* (Urbana: University of Illinois Press, 1987).

25. "A Complaint Regarding Conduct of Soldiers and Women in the Tent Adjoining to (Mary McHong)," in *Post Letters Received 5-28-06 to 9-22-06* (Record Group 393, part 5, vol. 87) Presidio, San Francisco; *New York Age* (May 3, 1906); Bronson, *The Earth Shook*; Gladys Hansen, *Denial of Disaster* (San Francisco: Cameron and Company, 1989).

26. Joseph Armstrong Baird, Jr., *Time's Wondrous Changes: San Francisco Architecture, 1776–1915* (San Francisco: California Historical Society, 1962); Harold Kirker, *California's Architectural Frontier: Style and Tradition in the Nineteenth Century* (Salt Lake City, Utah: Peregrine Smith, 1960); Richard Longstreth, *On the Edge of the World: Four Architects in San Francisco at the Turn of the Century* (New York: Architectural History Foundation, 1989).

27. *Historical Souvenir of San Francisco, Cal. With Views of Prominent Buildings, the Bay Islands, etc.* (San Francisco: C. P. Heininger, 1890); *San Francisco, Photogravures from Recent Negatives By the Albertype Company* (New York: A. Witteman, 1893); Charles Keeler, *San Francisco and Thereabout* (San Francisco: California Promotion Committee, 1902). Rodger Birt, et al., *San Francisco Album: Photographs by George Robinson Fardon* (San Francisco: Chronicle Books, 1999).

28. On the significant role classical antiquity played in American cultural life see Caroline Winterer, *The Culture of Classicism: Ancient Greece and Rome in American Intellectual Life, 1780–1910* (Baltimore: Johns Hopkins University Press, 2002).

29. Luigi Ficacci, *Giovanni Battista Piranesi, Selected Etchings* (Köln: Taschen, 2001).

30. Christopher, Woodward, *In Ruins* (New York: Pantheon Books, 2001), 89; the ruined City Hall was not the primary subject for either photographer but the structure's sheer bulk guaranteed its prominence in many photographs.

31. Eleanore F. Lewys, "A Dream of a Fair City," *Overland Monthly* 47, no. 3, 2nd Ser. (March 1906), 277–288; regarding

Phelan's "conflicted progressivism" see Philip J. Ethington, *The Public City, The Political Construction of Urban Life in San Francisco, 1850–1900*, 377–387; Phelan's enthusiastic embrace of the Burnham plan prior to the temblor and fire was noted in Gertrude Atherton, "San Francisco's Tragic Dawn," *Harper's Weekly* 50, no. 2 (May 12, 1906), 660; Judd Kahn discusses Phelan's political motives for opposing the comprehensive plan in *Imperial San Francisco*, 177–209.

32. Daniel H. Burnham, *Report on a Plan for San Francisco* (San Francisco: 1905); Phelan had said implementation of the plan would "take fifty years," Atherton, *San Francisco's Tragic Dawn*; *San Francisco Call*, May 23, 1906; *San Francisco Examiner*, March 11, 1909; *San Francisco Evening Post*, March 12, 1909; Charles A. Fracchia, "Make No Small Plans: The Burnham Plan for San Francisco," *The Argonaut* 15, no. 2 (Winter 2004), 46–61; also called "Goddess of Liberty," the figure's actual title was "Goddess of Progress" and was the work of sculptor Frances Marion Wells.

33. Cohen, "With a Camera in San Francisco"; Arthur Inkersley, "An Amateur's Experience of Earthquake and Fire," *Camera Craft* 12, no. 5 (June 1906), 199–200; Joseph Pennell, *San Francisco, The City of the Golden Gate* (Boston: LeRoy Phillips, n.d.), plate 4;

"Miss L. O. O'Bryan requests permission to make sketches of refugee camps . . . permission granted," *Post Letters Received* (May 31, 1906).

34. Louis J. Stellman, *The Vanished Ruin Era* (San Francisco: Paul Elder, 1910), 32–34; *Portola Festival Official Souvenir Program* (San Francisco: Portola Festival Committee, 1909); Frank M. Todd, *The Story of the Exposition* (New York: G. P. Putnam's Sons, 1921), 1, 43–46.

35. Pierre N. Beringer, "San Francisco's Wonder Year," *Overland Monthly* 49, no. 5, 2nd Ser. (May 1907), 386.

36. Edwin Emerson, Jr., "The Reconstruction of San Francisco," *Out West* 26, no. 3 (March 1907), 191–208; see also *Out West* 24, no. 6 (June 1906), 468–543.

37. Ethington, *The Public City*, 381; Kahn, *Imperial San Francisco*, 57–59.

38. Arthur Matthews, "Out of the Waste," *Philopolis* 3, no. 1 (October 25, 1908), 15.

39. Crittenden Van Wyck's album is in the collection of the San Francisco History Center, San Francisco Public Library.

40. See "Letter of Major Crittenden Van Wyck, D.D.S.," included with the album, n. 39; *San Francisco Chronicle*, April 18, 1941; *San Francisco Call-Bulletin*, April 18, 1941; Marilyn

Chase, *The Barbary Plague: The Black Death in Victorian San Francisco* (New York: Random House, 2003), 151–177; Kazin, *Barons of Labor*, 87–93; Walton Bean, *Boss Ruef's San Francisco: The Story of the Union Labor Party, Big Business, and the Graft Prosecution* (Berkeley: University of California Press, 1952), 250–252, 282–284.

41. Marjorie M. Dobkin, "A Twenty-Five Million Dollar Mirage," in Burton Benedict, *The Anthropology of World's Fairs*, (Berkeley: Scolar Press, 1983); Frank M. Todd, *The Story of the Exposition* (New York: G. P. Putnam's Sons, 1921), I, 43; *Portola Festival Official Souvenir Program* (San Francisco: Portola Festival Committee, 1909): *San Francisco Call*, Oct. 1–27, 1909 and Nov. 1, 2, 7, and 10, 1909.

42. San Francisco *Municipal Record*, 2, no. 42 (Oct. 21, 1909), 377, 386; *Portola Souvenir* is a "supplement" to San Francisco *Municipal Reports for the Fiscal Year 1908–1909, Ended June 30, 1909* (San Francisco, 1910), 66–67.

43. Will Irwin, *The City That Was—A Requiem of Old San Francisco* (New York: B. W. Huebsch, 1909), 7, 47.

INTRODUCTION TO THE PHOTOGRAPHS

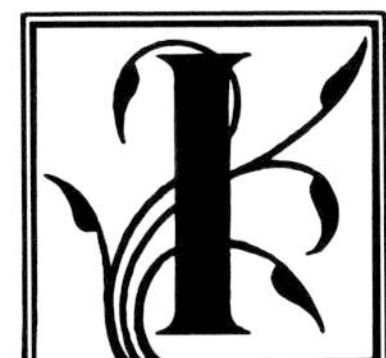

t is a challenging task to present in a logically ordered manner a disparate group of photographs made over a three-year period united by topic: earthquake, fire, and recovery, but taken at various times in various places under varying circumstances by a wide variety of cameramen. There are a number of ways in which such a collection of views could be organized. The most obvious, but in some respects the most artificial, is by chronology since temporal sequence is a traditional technique for arranging documents in a comprehensible form. Here, such categories as "earthquake," "fire," "immediate aftermath," and "long-term recovery" might be employed. Another method for organization is by geographical setting, though focusing on neighborhoods in San Francisco over a three-year period involves the problems of repetition and of failing to give a sense of the city's overall response to the disaster. A third approach to organizing a large number of images with thematic connections is to categorize them by the type of photographer who shot each image. Such categories as journalistic photographer whose primary aim is to inform, professional/studio photographer whose ideal is to captivate, and art photographer whose goal is to use the particulars of disaster and recovery to make transcendent visual statements about human nature, the impermanence of life, or the fall from high estate might serve to bring order to so varied a group of images. However, such a typology of photographers faces the inevitable problem of overlap since the lines among journalistic, professional, and artistic photography are often hard to draw, and some photographs are certainly made with more than one intention in mind. Again, a large group of thematically related images might be arranged by the particular photographers who made them. The problem here is twofold. First, some of the views included in this collection were taken by unidentified cameramen working for large studios. Second, this approach has built-in imbalances since some of the photographers were systematic in their memorializing of the disaster and its aftermath, while others, often the amateurs, were more idiosyncratic in their recording of events. A final method of organization is by the relation of the camera to the objects in the field of vision. One group of images would include highly intimate photographs of particular sites, another would include scenes in which the viewer's eye is drawn from foreground to middle distances, and a third would include panoramic views made from great distances.

While each of these methods of organization has its advantages and drawbacks, we thought it best, in the interest of clarity, that the descriptions of the photographs in this collection be based primarily on chronological sequence. However, the content of each description includes not only the identification of location and date or approximate time of the image but also abundant observations on visual composition, details of historical interest, local history in a more general sense, architectural style and architects, and the intentions of various photographers in creating particular views. Such an amalgam is designed both to describe what is seen in a particular image and to explain its significance in broader cultural terms. Through this approach we hope to demonstrate the trajectory from initial shock and dismay to self-confident resurrection experienced by the people of San Francisco during the three years following April 18, 1906.

Marvin Nathan
Berkeley, California

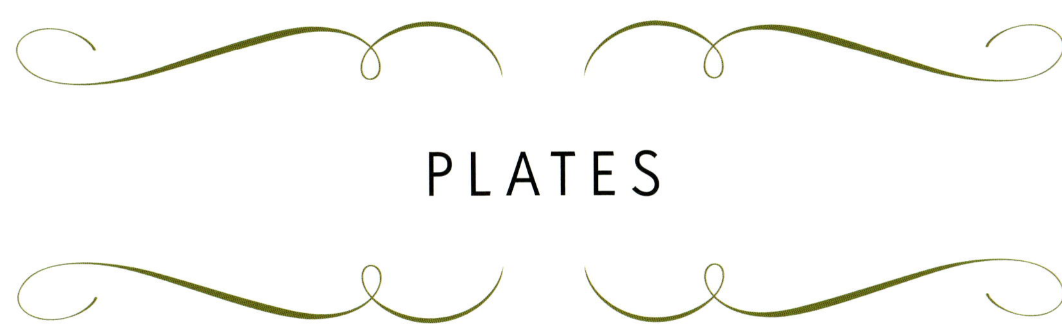

PLATES

In the weeks after the disaster an anonymous photographer shot this stereographic view of Union Street looking east between Pierce and Steiner Streets in the northern sector of the city. Since the fire did not burn this area of San Francisco, there are visible no charred remains of clapboarded Victorian homes. However, earthquake damage is apparent enough. Both the uplifted pavement in the foreground and the gaping hole on the second story corner of the house at the middle right provide evidence of the tremor's force. Despite the damage upon which the camera dwells, houses, telephone poles and the street in the distance seem relatively intact. This image is a good example of how photography, like all other media, can create a sense of drama by selection and emphasis. The sign boards in the view also suggest how, by 1906, satellite shopping centers had grown in less densely populated areas of the city.

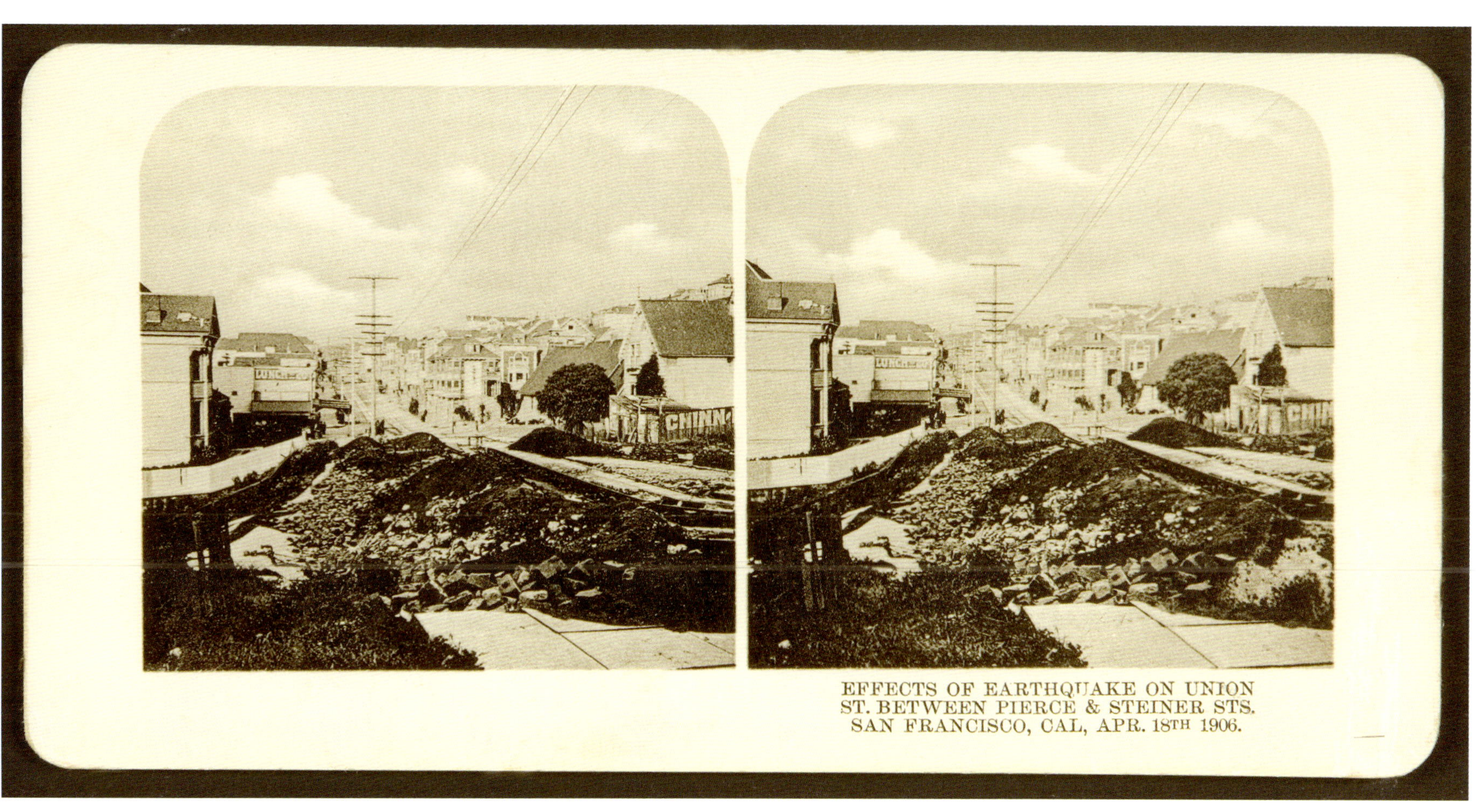

EFFECTS OF EARTHQUAKE ON UNION
ST. BETWEEN PIERCE & STEINER STS.
SAN FRANCISCO, CAL, APR. 18TH 1906.

PLATE 2. Unattributed, *Looking South From 17th and Howard Sts.,* 1906, photolithograph (private collection)

The Mission District, which was built on loose alluvial earth formations crossed by underground creeks, suffered mightily from the shaking both in the areas that were burned and those left untouched by fire. In this stereographic view a cameraman stands on Seventeenth Street looking south along Howard (now South Van Ness) capturing much of the earthquake damage over a two-block area. Along with the dramatically captured road destruction in the foreground, a row of Italianate, Eastlake and Queen Anne houses displays a boarded up façade in the foreground and crazily tilted structures along the receding street. Why some structures withstood the tremor and other nearby structures failed was a question that government and other engineers puzzled over for years after 1906.

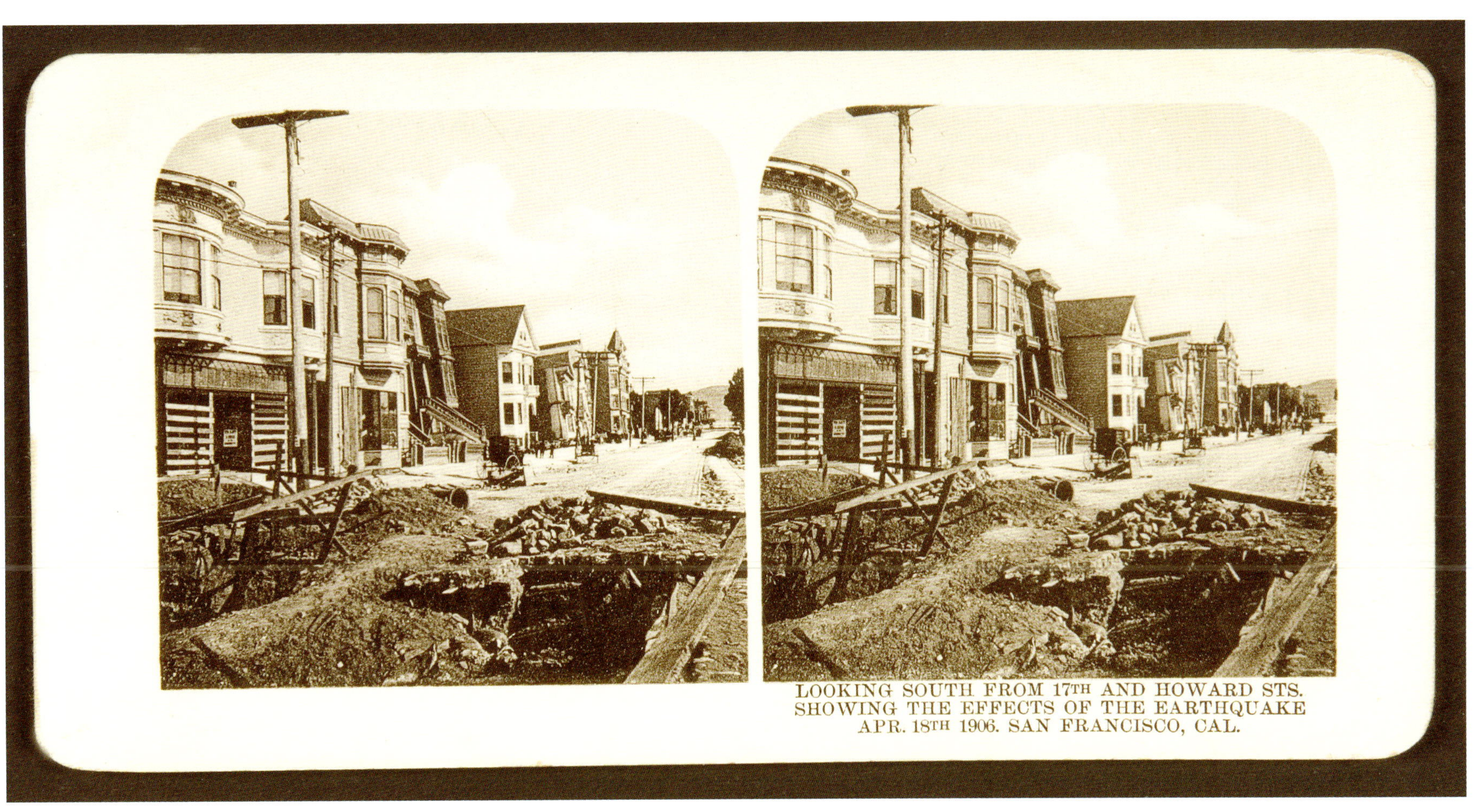

LOOKING SOUTH FROM 17TH AND HOWARD STS.
SHOWING THE EFFECTS OF THE EARTHQUAKE
APR. 18TH 1906. SAN FRANCISCO, CAL.

What is perhaps most striking in this stereographic view of the destroyed City Hall is the way in which the anonymous photographer evokes the emptiness surrounding the ruins in the destroyed sections of San Francisco. Innumerable photos were taken of this structure, its shaken and charred bulk looming large over the cityscape. But this image introduces the viewer not only to the demise of a great public building—which was barely seven years old and had cost tax payers millions of dollars in budget overruns—but also to the desolation that San Franciscans like those stylishly dressed, derby hatted ones depicted here had to face in the weeks, months, and even years after the great disaster. In the distance at the middle right of the view stand the ruins of the Hall of Records where fire destroyed the birth certificates and other vital statistics of two generations of the local citizenry. This combination of destruction, desolation, and loss in the city's precincts must certainly have taken a heavy psychological toll. The ability of San Franciscans to overcome this pall and begin the swift rebuilding of their community is a testament to their pride of place, civic energy, and economic optimism.

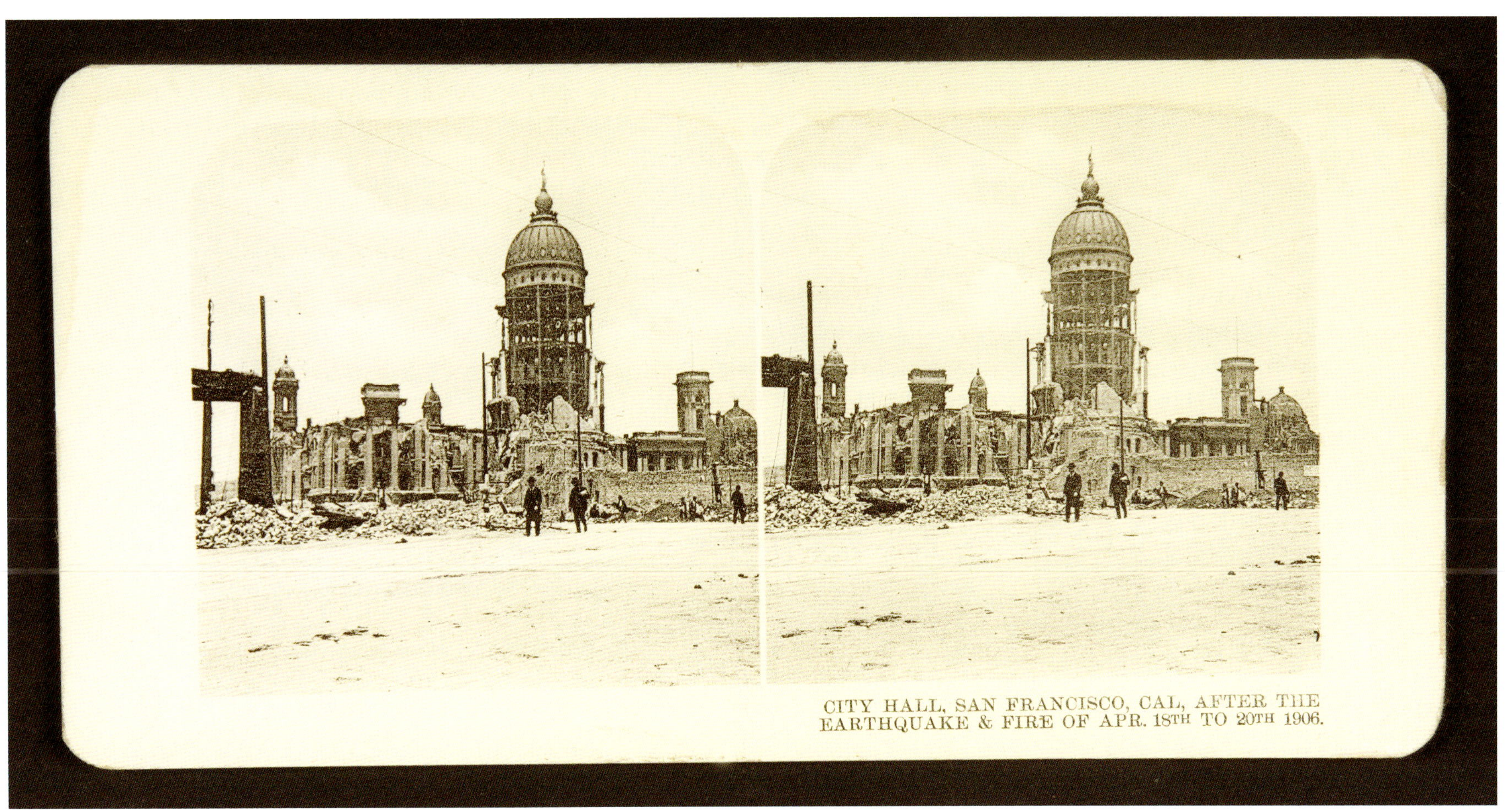
CITY HALL, SAN FRANCISCO, CAL, AFTER THE
EARTHQUAKE & FIRE OF APR. 18TH TO 20TH 1906.

This rare photograph of the interior of a house immediately after the earthquake was made by an unidentified photographer (whose own residence this may be). This rather reportorial view shows substantial damage to the front parlor caused by ground uplift. Also pictured are such Victorian domestic staples as gas fixtures, decorated fireplace surround and mantle, pocket doors, and bay windows. At the front left in the family parlor stands a status symbol in many Victorian middle class homes, the upright piano.

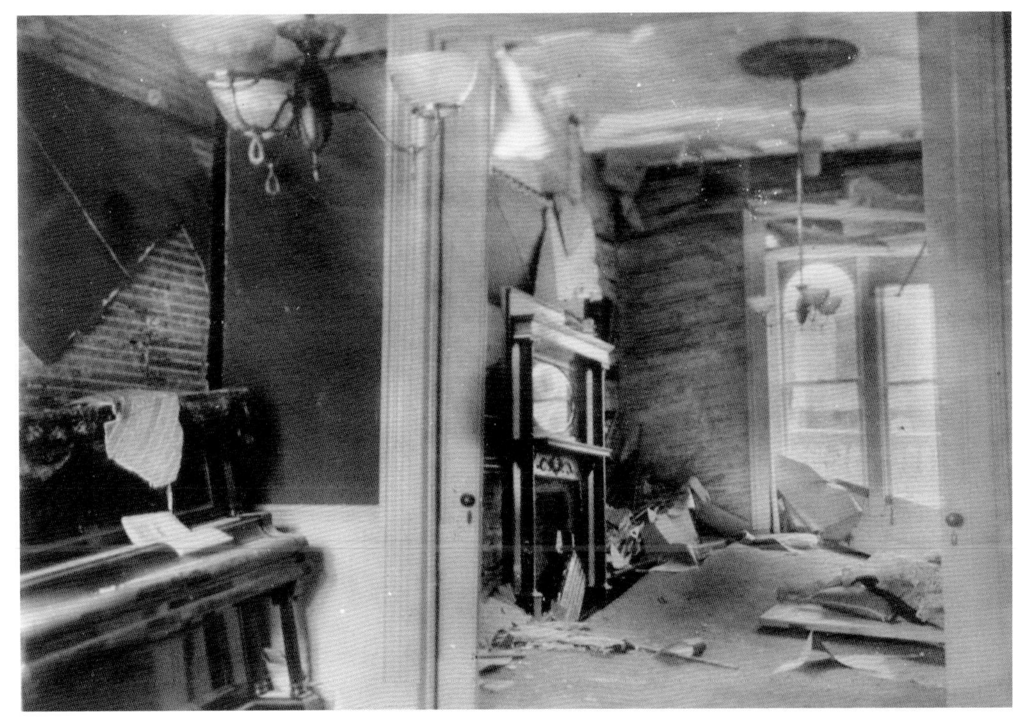

PLATE 5. Willard E. Worden, *Fire Looking from Nob Hill,* 1906, gelatin silver print (private collection)

Willard Worden, a local commercial photographer, made this panoramic view of downtown San Francisco from a site near his house on Nob Hill. The drama of this image is apparent as the cloud of smoke south of Market Street hovers menacingly behind the still largely undamaged business buildings just to the north. Within two days most of the structures in this photograph would stand gutted or be entirely destroyed by flames. Several of San Francisco's important buildings are shown in this view. Near the center of the picture is the Call Building, the tallest skyscraper in the city, designed by the Reid Brothers and dating from 1898. Across Market from the Call is William Curlett's Mutual Savings Bank Building dating from 1902. To the right of the Call Building is the still unfinished Humboldt Bank Building designed by Meyer and O'Brien. All three of these structures still stand along Market Street, though in somewhat altered form. In the middle of the photograph is William Patton's onion-domed Temple Emanu-El at 450 Sutter Street built in 1865. Note the unfinished steel-reinforced commercial structure at the right center. Modern construction techniques allowed many of San Francisco's newer buildings to withstand the massive earthquake with relatively minor damage. This view typifies so-called "smoke pictures," photographs taken of the city ablaze which were popular commodities both locally and nationwide.

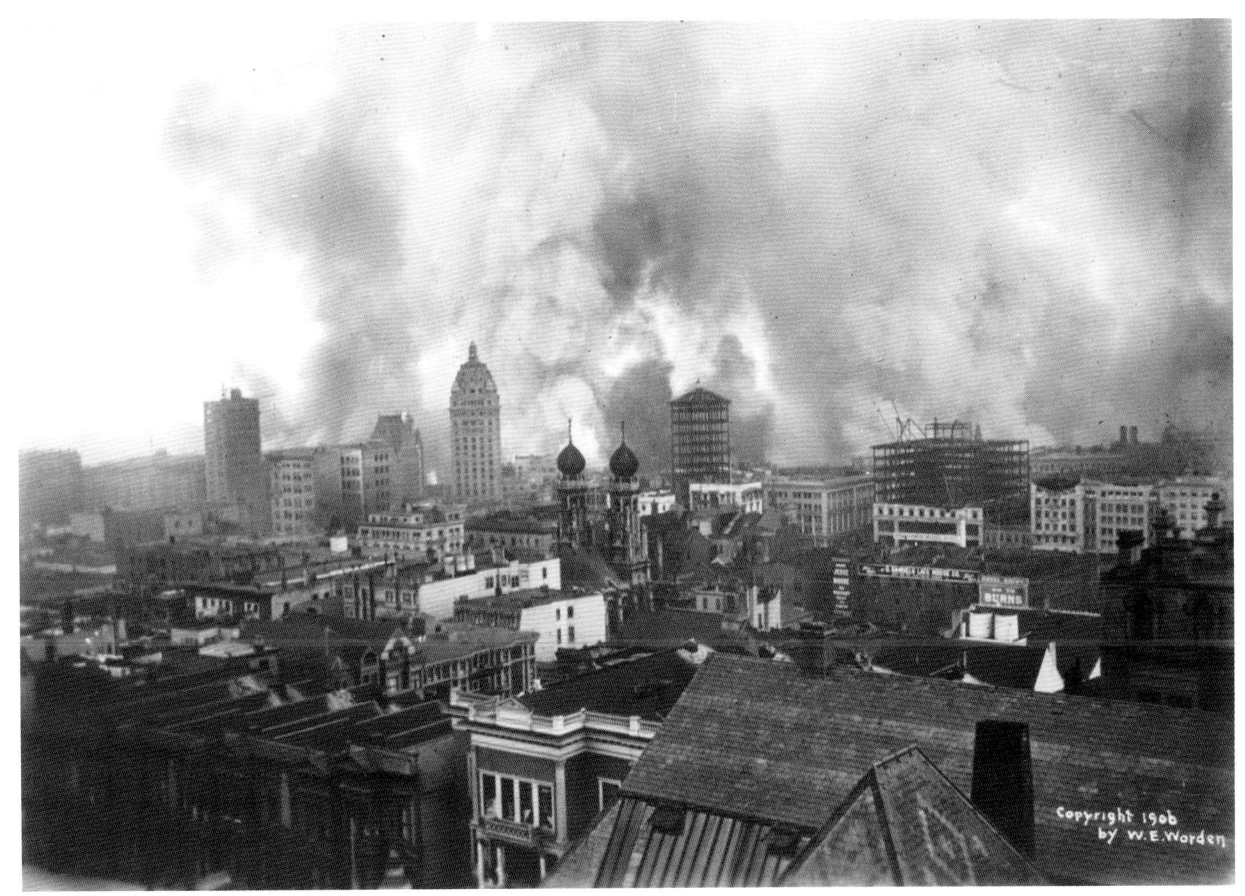

Copyright 1906
by W.E.Worden

This is Worden's view of downtown taken from O'Farrell Street looking southeast toward Market. Again we see the Call Building with clouds of smoke approaching it. Pictured along Geary are structures in both neoclassical and Victorian style. Other points of interest are the rubble created by the earthquake, the undisturbed cable car tracks, the Masonic order sign, and the transfixed crowd along the street. There seems no attempt here on Worden's part to amaze or moralize, but rather he merely reports a great event as it unfolds before the spectators' eyes.

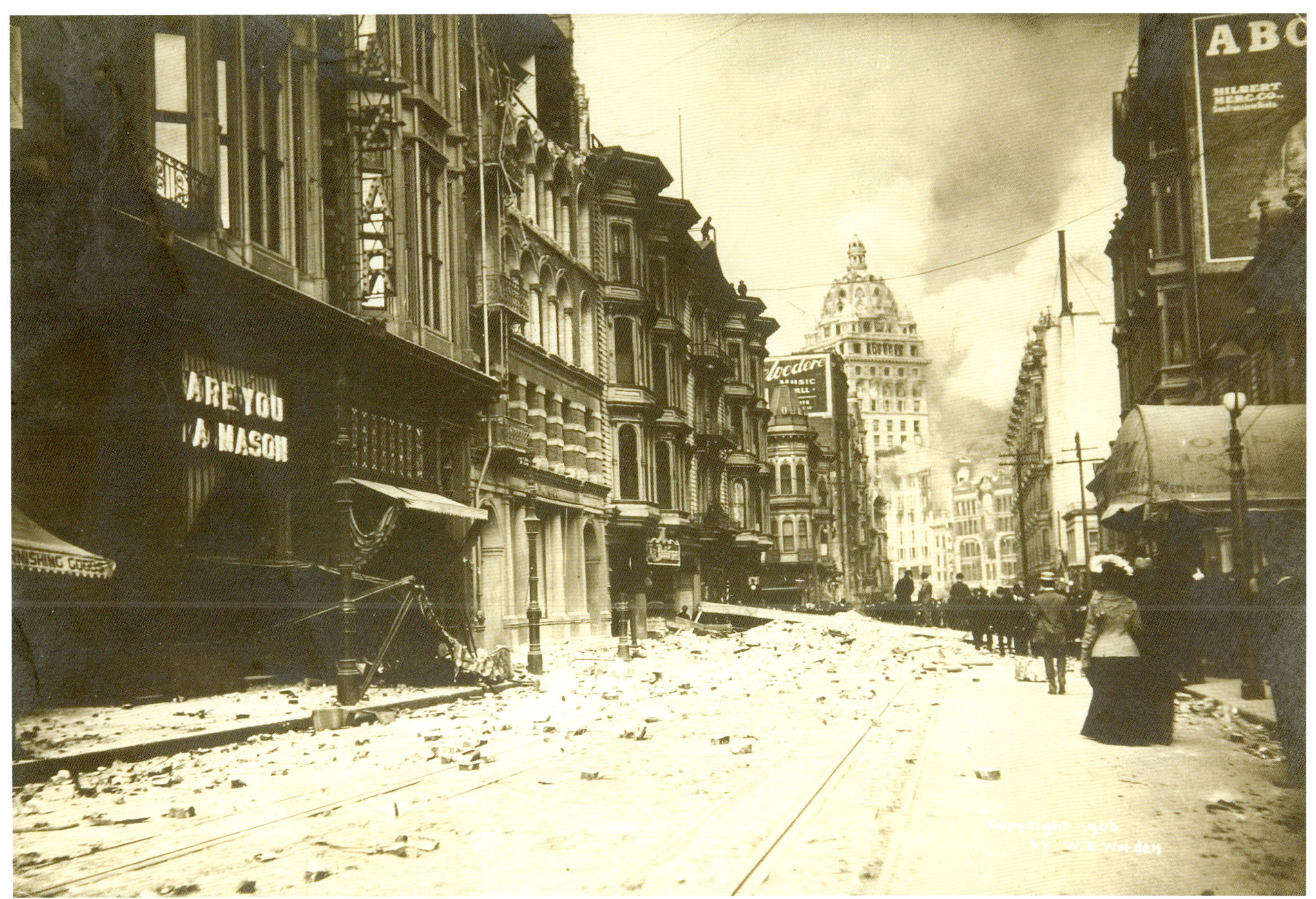
ARE YOU
A MASON
ABC
HILBERT
MERC. CO.
Floeders
MUSIC HALL
Copyright 1906
by R. E. Warden

PLATE 7. Unattributed, *Untitled,* 1906, gelatin silver print (private collection)

Looking eastward down Pine Street from just below the crest of Nob Hill, an unnamed photographer aims his camera past the gold rush landmark old St. Mary's church at California and Dupont (Grant Avenue since 1908) to record the fire's progress northward along the Embarcadero. Old St. Mary's, the first great religious building in American San Francisco, survived the disaster but was seriously damaged inside and out by both earthquake and fire. Just above the church's roofline and to the left of the bell tower appears another gold rush–era building, the famed Montgomery Block at Montgomery and Washington Streets where it stood until 1959. Immediately above it towers the multichimneyed Appraiser's Building, which survived the fire. To the far right the north wing of the towering Merchant's Exchange is visible. It too was gutted by fire. The hilly geography immediately surrounding the bowl-like topography of downtown San Francisco permitted photographers of the disaster to compose views containing remarkable panoramic sweep without losing the intimate details of particular buildings being or about to be engulfed.

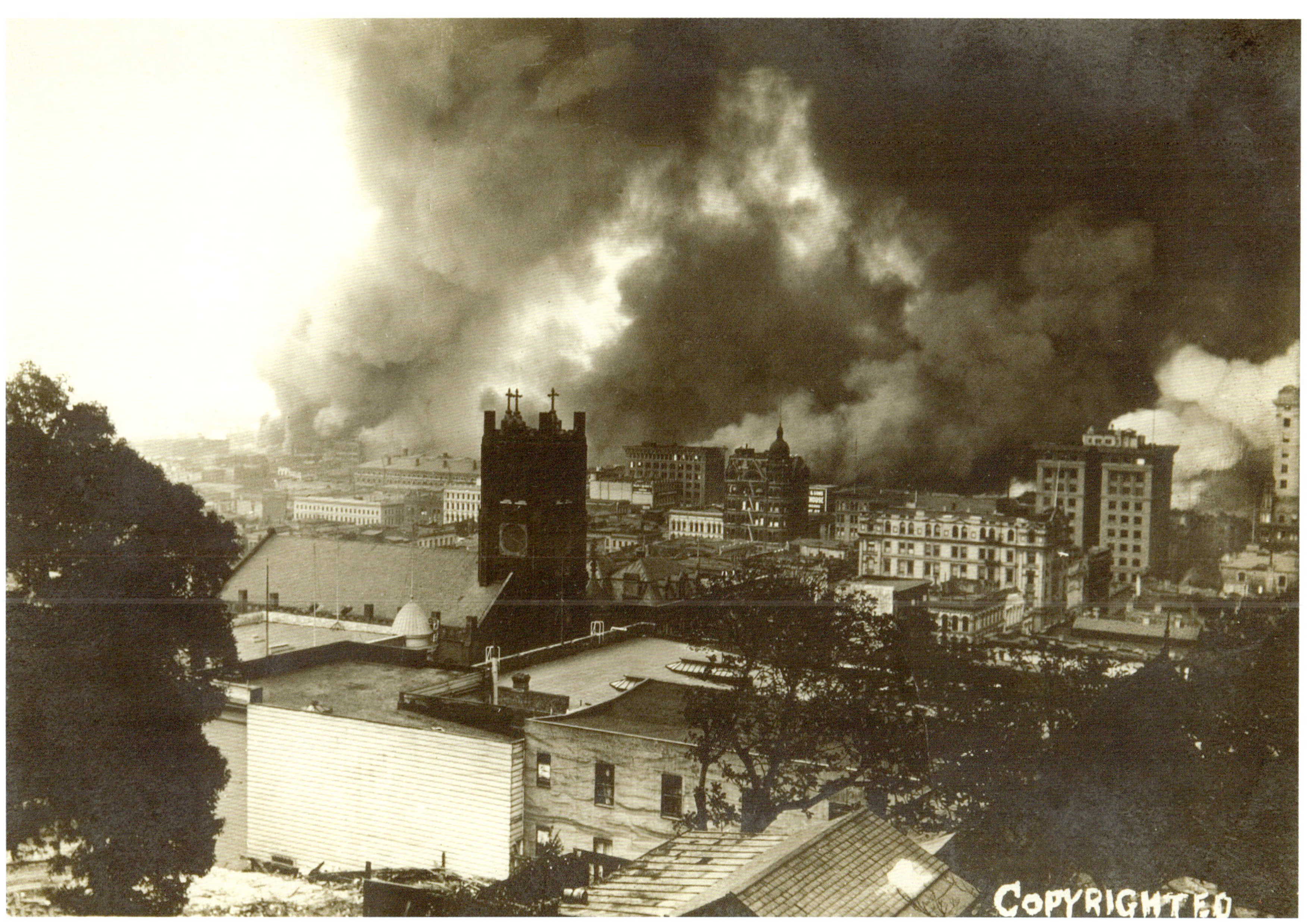

COPYRIGHTED

PLATE 8. Willard E. Worden, *Untitled,* 1906, gelatin silver print (Wells Fargo History Dept., San Francisco)

During the days of fire Willard Worden spent a good deal of time photographing scenes of destruction on or around Market Street. In this view he stands near Battery Street pointing his camera toward the south side of Market Street at two Baroque Revival business buildings, one already decimated by fire, the other in the midst of conflagration. All of the normalities of the 1906 city are shown: cobblestone streets, cable car tracks, decorative gas lamp street lights, a dozen well-dressed men, and masonry buildings designed in the typical revivalist styles of the late nineteenth and early twentieth centuries. Except this is not a normal San Francisco morning. The city which had seemed so solid and indestructible a day or two earlier was now being demolished before the unbelieving eyes of local citizens. Worden's view captures the befuddlement of the witnesses, their desire to seek social contact in order to share news and anxieties, and the desultory nature of behavior in the face of frightening, unexpected change in human circumstances. The figures here are both witnesses to and actors in the great drama unfolding around them.

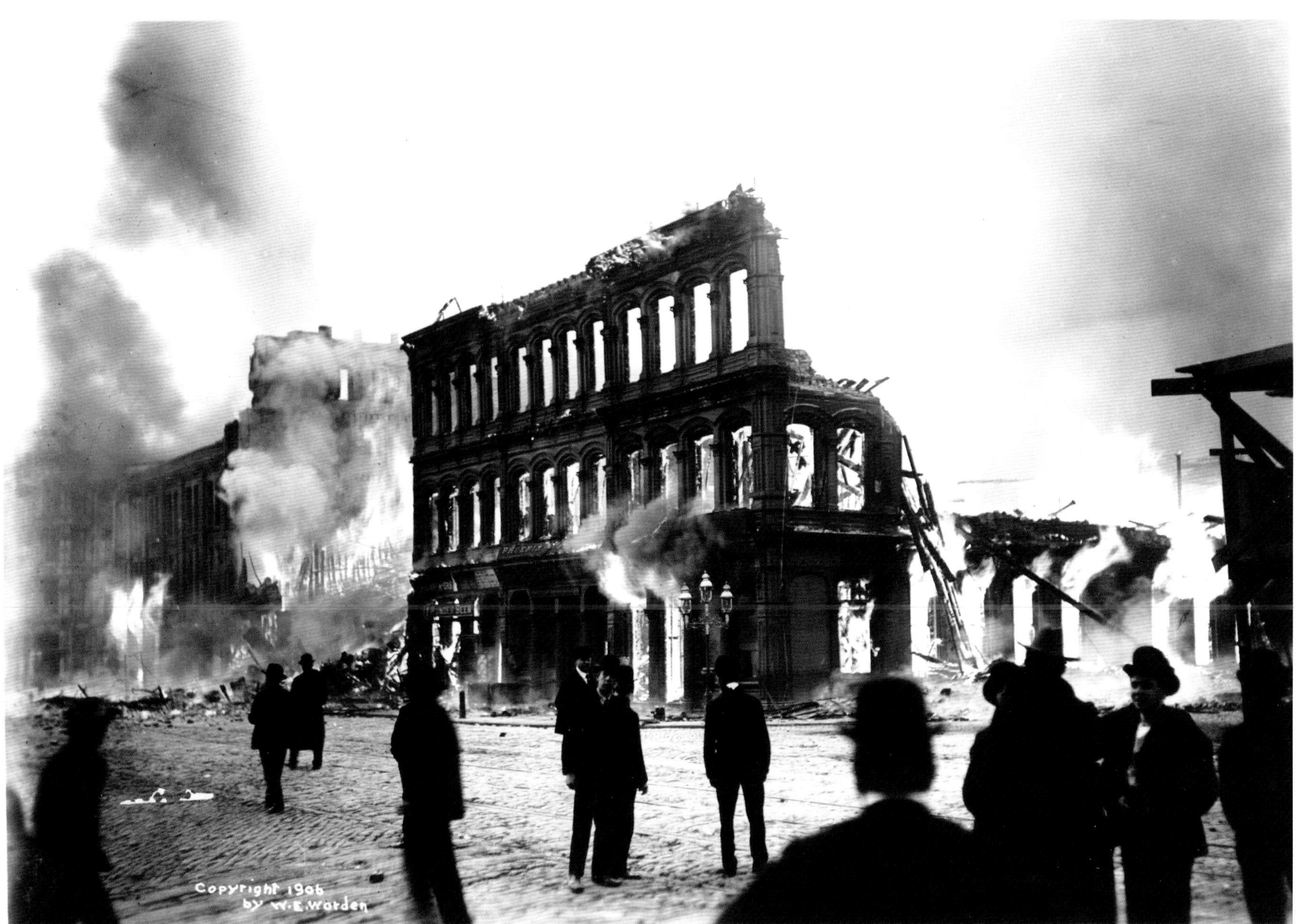

Copyright 1906
by W.E.Worden

PLATE 9. Willard E. Worden, *Untitled,* 1906, gelatin silver print (Wells Fargo History Dept., San Francisco)

In this view Worden stands on the northeast corner of Montgomery Street with his camera trained west along Market toward the as yet undamaged Palace Hotel and the Call Building whose windows exhale large quantities of white smoke. Although the two great structures tower over the scene, Worden's attention seems caught by the casual demeanor of the police and civilians in the foreground, implying that neither they nor the photographer as yet understand the dire consequences for the city of what they are witnessing. Note the relatively undamaged condition of the Market Street roadway and street lamps and the heavy reliance of the local authorities on horses as the primary means of transportation while patrolling the center of San Francisco. However, in the days and weeks after the earthquake and fire the clear advantages of the automobile, its speed and agility, along with the deaths of an estimated 15,000 horses, markedly helped San Francisco progress from the equine to the internal combustion era.

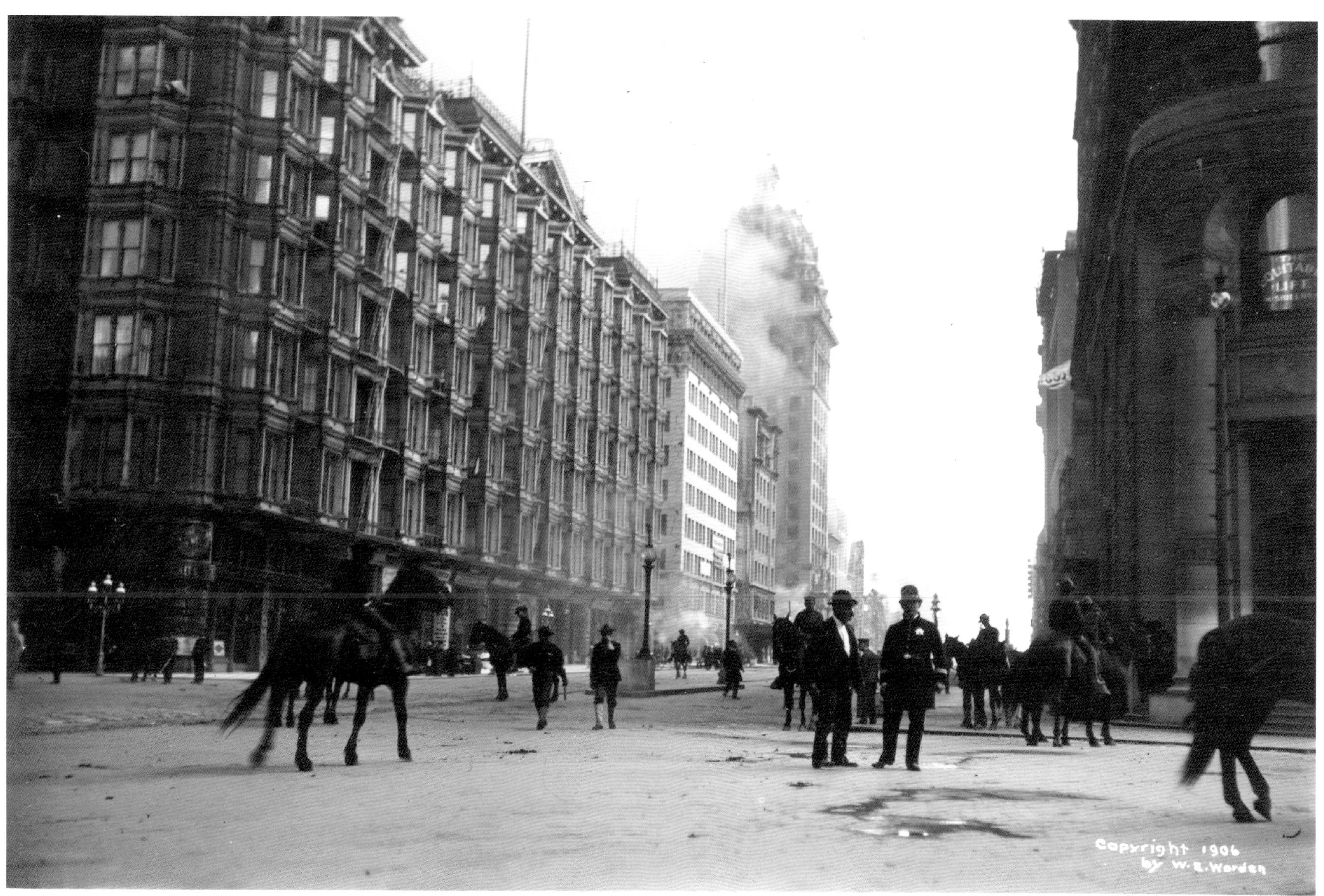

Copyright 1906
by W. E. Worden

PLATE 10. Pillsbury Picture Company, *Crossley Building on Fire,* 1906, gelatin silver print (private collection)

An unnamed photographer working for the Pillsbury Picture Company captured the swirling conflagration as it raced west up Mission Street and devoured the old Crossley Building at the corner of New Montgomery. The flames would soon advance upon Wade's Opera House and the Palace Hotel, each within a block and a half of the Crossley. The (anonymous) photographer cleverly uses the billowing cloud of smoke to express both the immediacy and the immensity of the disaster that had befallen the city. The first Crossley Building shown here was owned by the Spreckels family and built by the noted architectural firm of Meyer and O'Brien. A new Crossley Building designed by Mel I. Schwartz was erected on the same site in 1907.

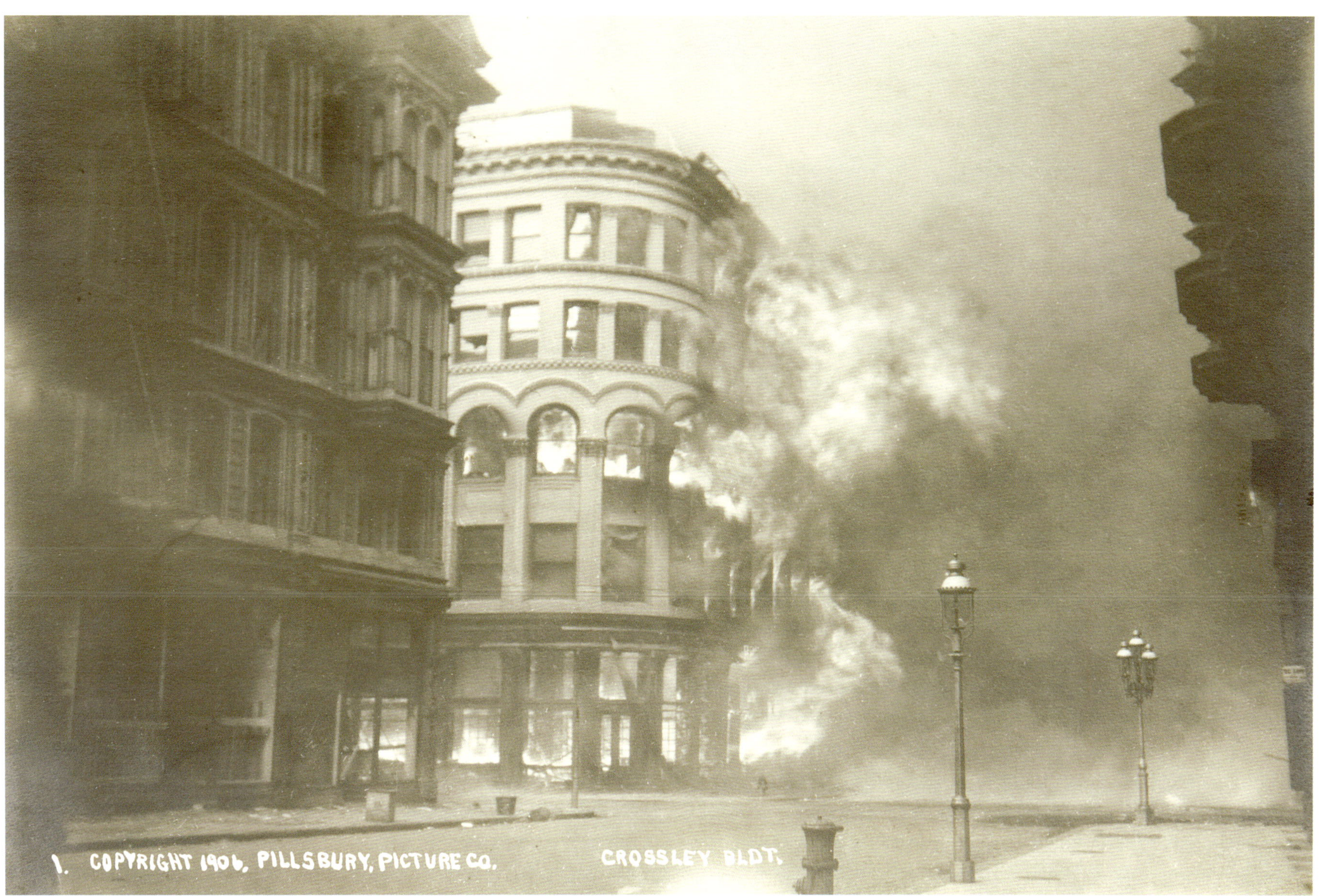
1. COPYRIGHT 1906, PILLSBURY, PICTURE CO.   CROSSLEY BLDT.

This Pillsbury Picture view, taken along Sansome Street, depicts not only the Dantean inferno and the miasma of smoke that would engulf the whole downtown area by April 19 but also the manner in which many of San Francisco's largest commercial structures would be destroyed by flames eating away at their interiors floor by floor. Photographs such as this one seem designed to evoke emotion, not to be journalistically specific or to transport the viewer to thoughts about divine destiny and the impermanence of things human.

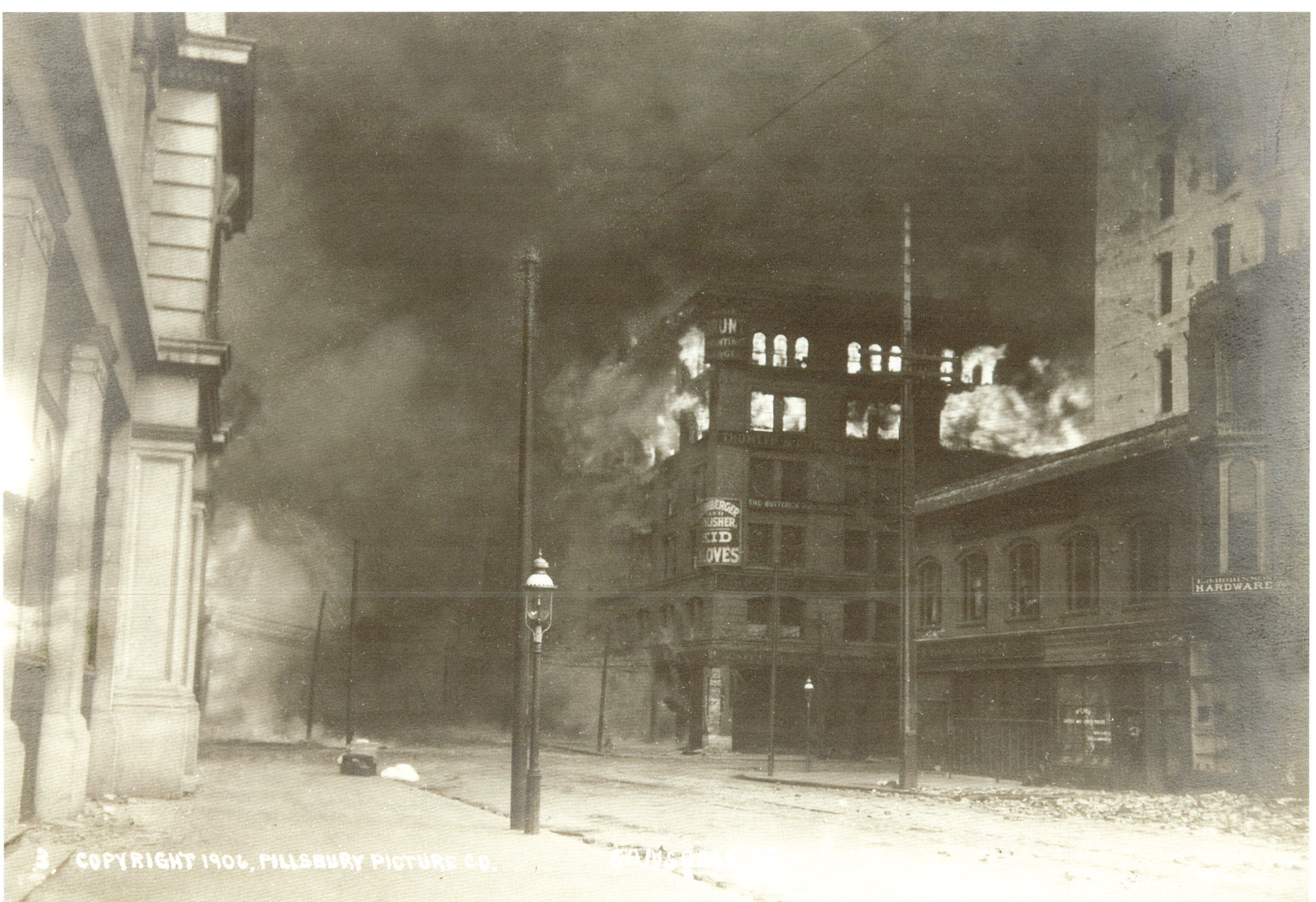
3. COPYRIGHT 1906, PILLSBURY PICTURE CO.

PLATE 12. Pillsbury Picture Company, *Palace Hotel on Fire,* 1906, gelatin silver print (private collection)

This view, photographed by a Pillsbury Picture Company cameraman, was taken from Montgomery Street just above Market and shows the destruction of two great buildings from the 1870s. The smaller Second Empire-style building on the left is banking millionaire William Ralston's Grand Hotel, a hostelry representing San Francisco's elegance as a major American city during the Victorian era. The larger structure on the right is Ralston's world renowned Palace Hotel, even as late as 1906 the most famous building in the city. Ironically the Palace, with its 800 rooms and 700 bay windows, had been built with a huge water reservoir in its basement to protect it from fires such as the one that now consumed it. Sadly, the water supply was spent before the worst of the flames reached the hotel. The three firemen manning a waterless hose are an eloquent symbol of the magnitude of the city's predicament. Few San Franciscans could view this photograph and not be moved by what it represented about a lost past and an uncertain future.

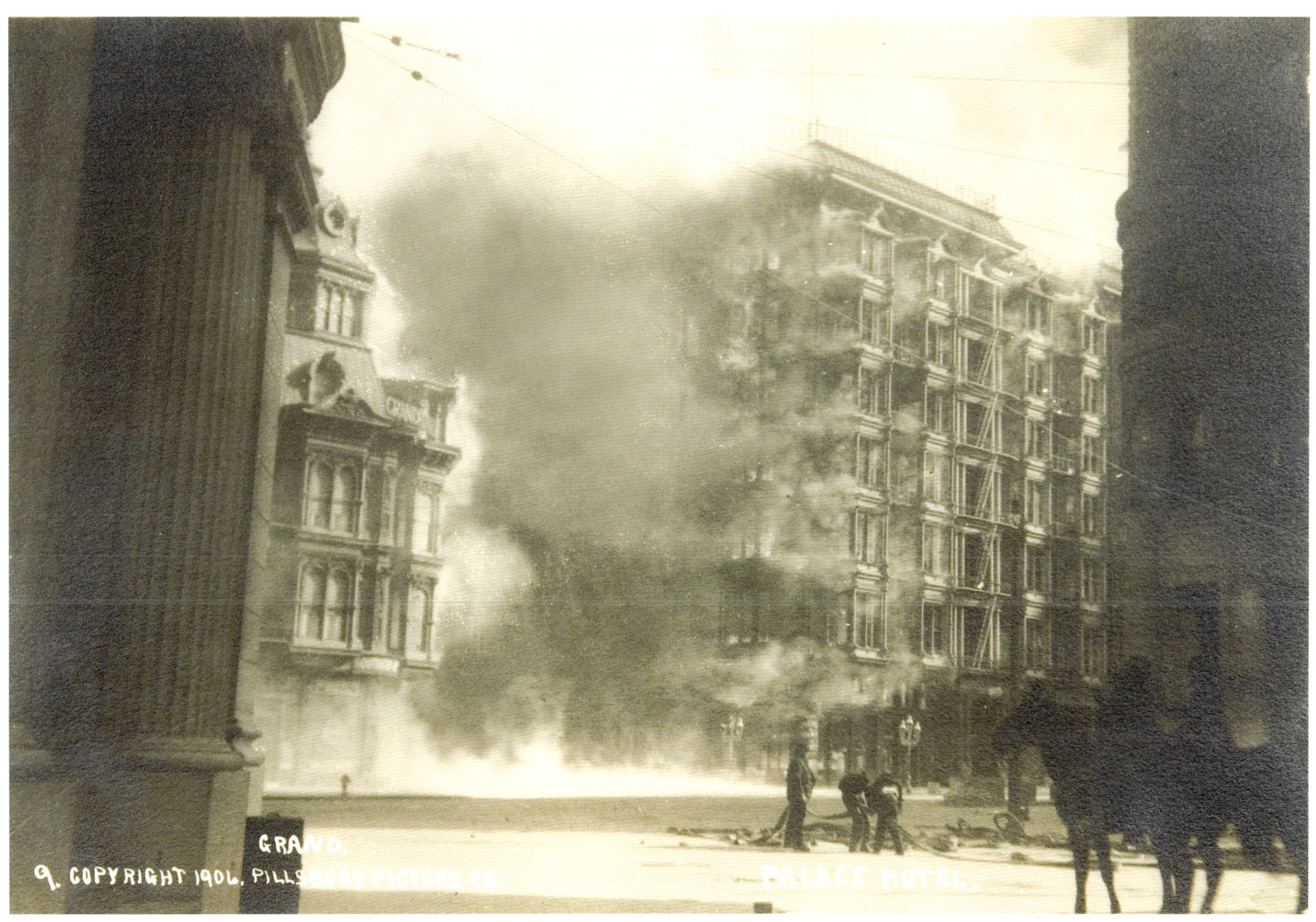
GRAND
9. COPYRIGHT 1906, PILLSBURY PICTURE CO.
PALACE HOTEL.

This award-winning view of crowds on Sacramento Street atop Nob Hill watching the encroaching wall of smoke and flame moving ever nearer was made by Arnold Genthe, perhaps the most gifted and best known professional photographer working in San Francisco at the time of the earthquake and fire. This dramatic scene of citizens, flanked by earthquake damaged buildings, intently watching the destruction of the heart of their city like spectators at a great theatrical performance, is laden with both sadness and irony, for only a short time later their position of vantage would be reduced to ashes and rubble by the onrushing flames. Grandiose drama and panoramic visions of ensuing Armageddon characterize many of Genthe's photographs in the three days of earthquake and fire. In this view he uses the natural perspective created by Sacramento Street's southeastern descent toward Market to focus the viewer's eye, only to have the huge plume of smoke obliterate any sense of a clearly defined vanishing point. This is a case of catastrophe obliterating perspective, which, metaphorically speaking, was what was occurring in San Francisco at this moment of crisis. Note that the cable car tracks and the wooden buildings have endured the quake with no apparent serious damage whereas the unreinforced masonry building on the far right has sprung its front wall.

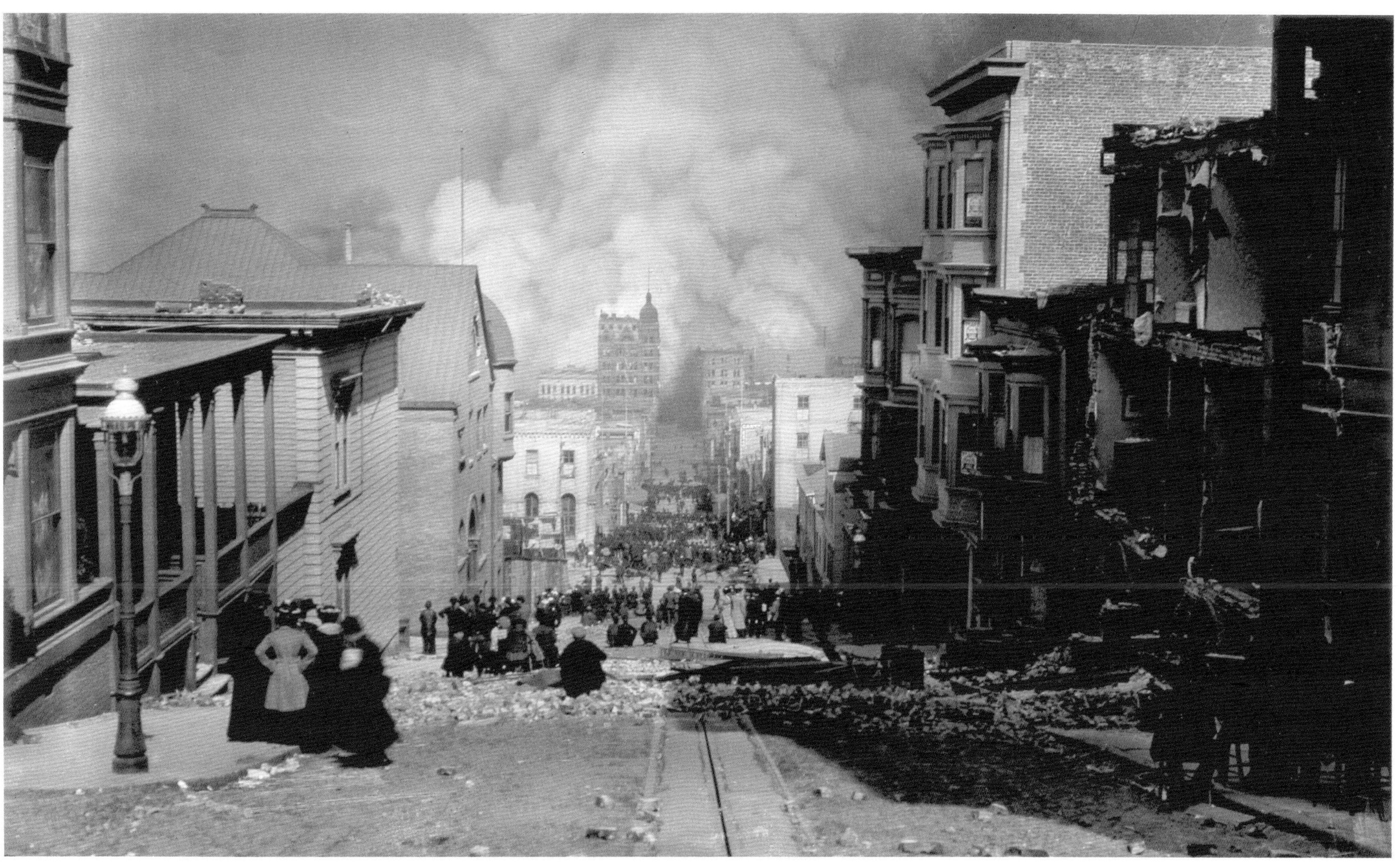

Genthe is best known for his images of San Francisco's pre-earthquake Chinatown. An intellectual and a liberal, the German-born photographer exhibits in this view of Sacramento Street above Grant Avenue, near the heart of Chinatown, a continuing interest in the city's multicultural society, this time at a moment of crisis. Chinese, African-American and Caucasian individuals, all in similar attire and all seemingly waiting for some resolution to the catastrophe pictured in the background, exhibit the democratizing effects a disaster imposes on an urban population. On April 28 the *San Francisco Chronicle* asserted that "Society is on the ground, face to face, jowl to jowl. Every artificial barrier is swept away." The brilliance of Genthe's presentation of this theme is here apparent in the subtlety of its expression.

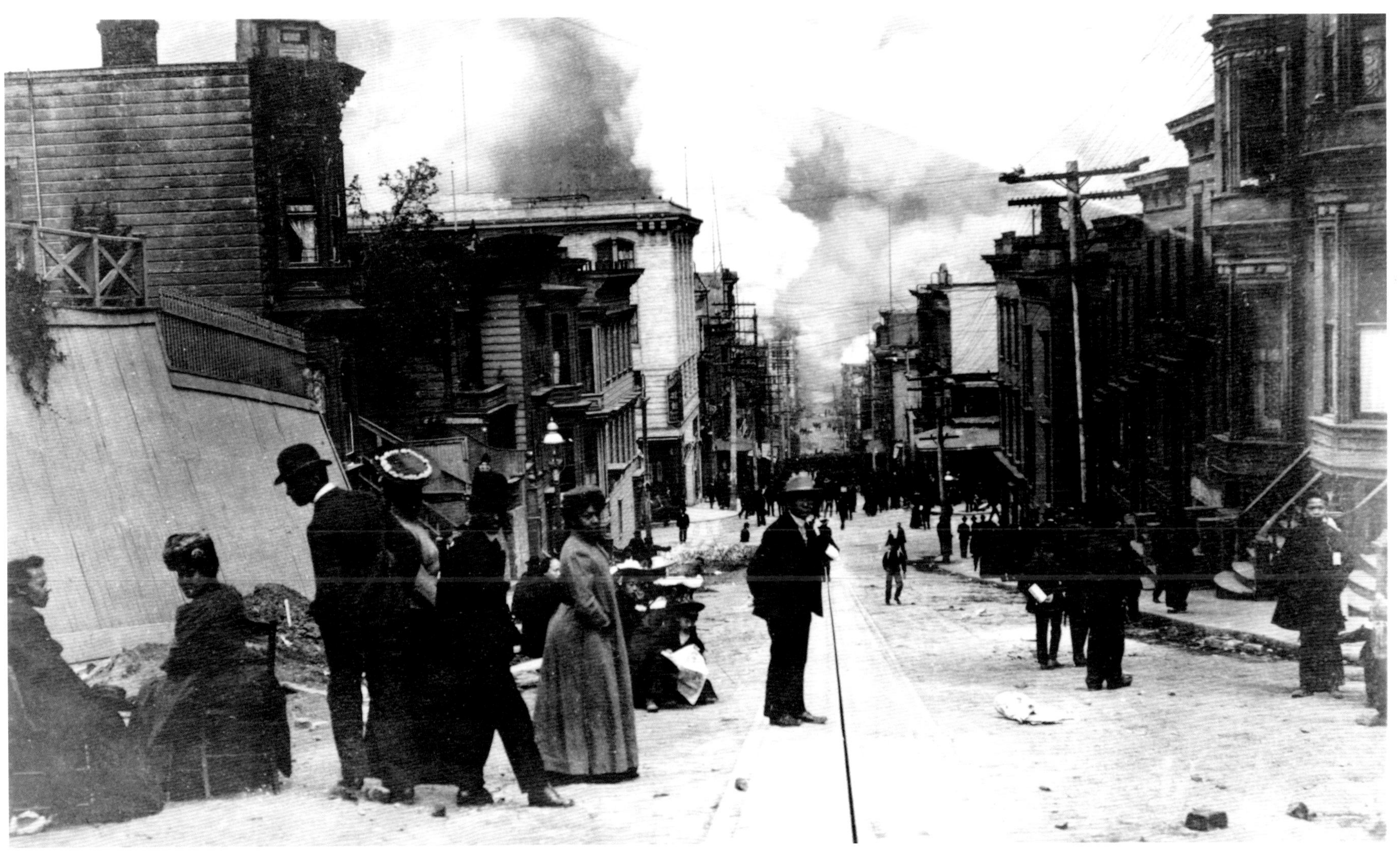

This anonymously taken view, shot at the height of the fire, leads the viewer's eye from the male figure in the foreground through various planes eventually disappearing into the obliterating smoke covering downtown. We stand at the highest point in Pacific Heights, 378 feet up in Lafayette Square with a group of informally dressed San Franciscans viewing the disaster to the southeast from a makeshift camp of tents, sleeping rolls, blankets, trunks and baby carriages. Lafayette Square is located just west of Van Ness bordered by Gough, Sacramento, Laguna, and Washington Streets. The drama of this image lies in our shared sense of the inexorable movement of the conflagration to the north and west of the city and the anxiety of those fleeing their homes to arrive at a safe distance from the holocaust. As it would turn out, the flames were extinguished just one block short of where these figures are standing. Both the menace of the moment and the dramatic disruption of normal life are significant themes seen in this image. Yet despite the dire circumstances, the hands on hips poses of the figures in the foreground imply a resolute attitude about coping with the disaster on the part of the people depicted and, perhaps, of the photographer himself.

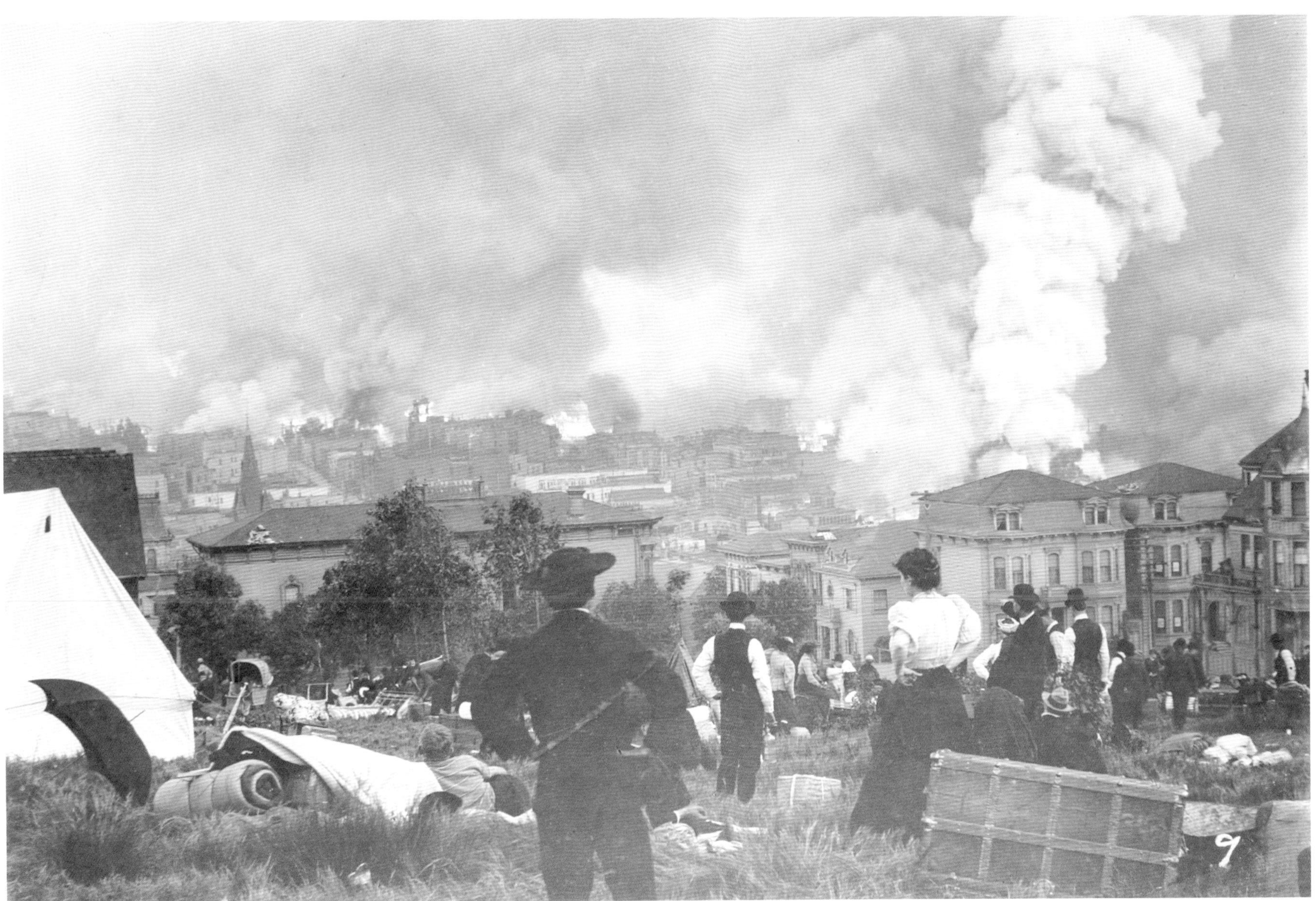

PLATE 16. Louis J. Stellman, *Untitled,* 1910, photogravure (private collection)

This image of men evacuating the endangered areas of the city under military order reveals the two sides of the photographer Stellman's visual vocabulary. It is clear that when he took this picture, on the day of or day after the disaster, his journalistic background led him to train his lens on detailed reportage of the unfolding catastrophe—reflected not only in badly damaged buildings and a rubble strewn landscape but also in the faces and postures of the fleeing men who possessed little beside what they could carry in hand. Yet, when he published his collection of earthquake photographs four years later in 1910, his sharp-eyed journalistic version has been manipulated in the darkroom and has become a pictorialist work utilizing soft focus and muted tones reminiscent of nineteenth-century Tonalist painting. The aim of such techniques was certainly to deepen the emotional resonances of human calamity less by the graphic presentation of a specific historical moment than by the evocation of a pervasive atmosphere of melancholy and resignation.

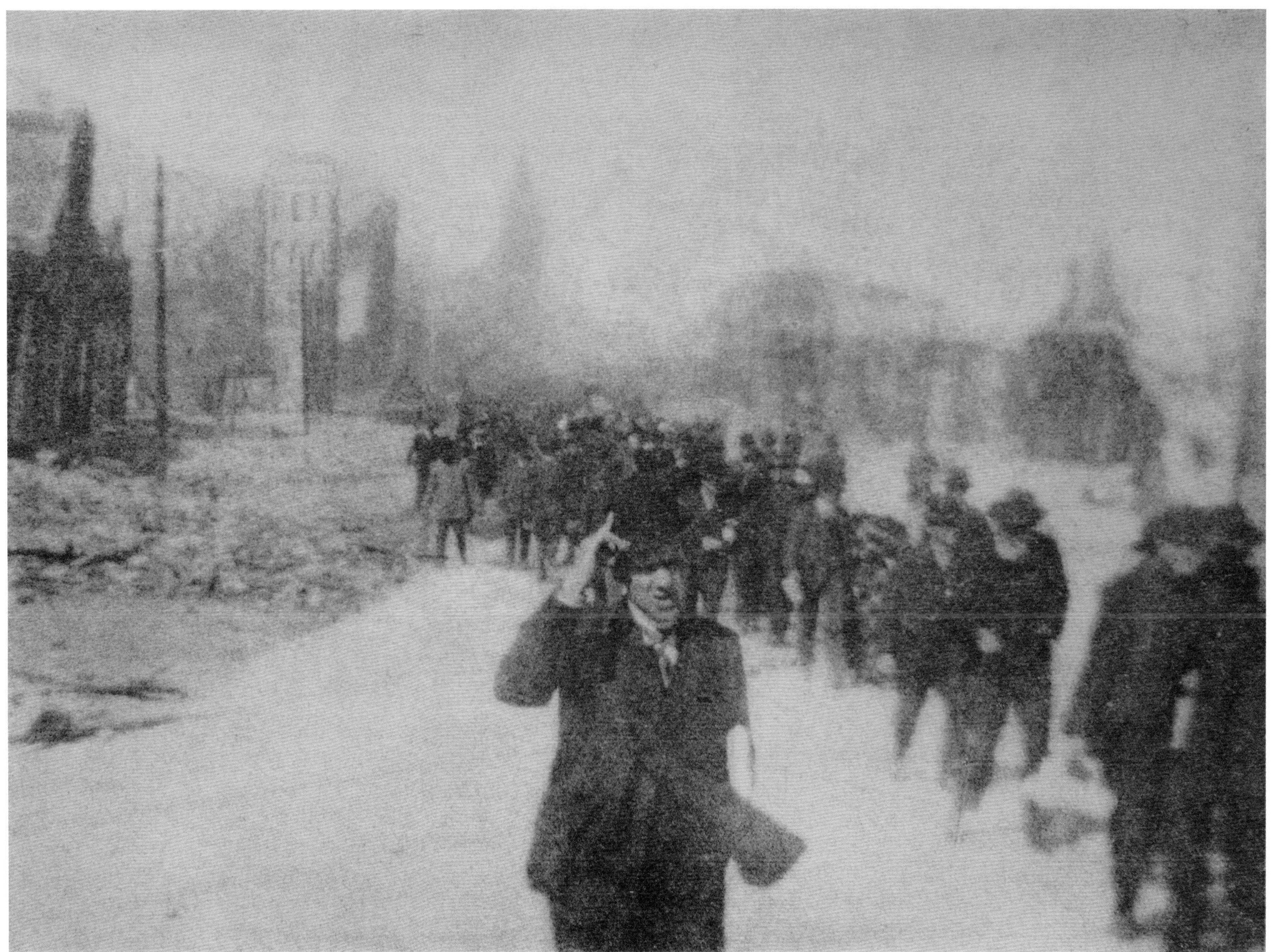

Professional photographer James D. Givens made a concerted effort in the weeks after the earthquake and fire to create a tapestry of views revealing the condition of San Francisco as it staggered toward recovery. Here is the first of Givens's panoramic views of the city's central area, this one scanning the downtown section in a southeastern direction from a vantage point atop what appears to be the Flood Mansion on Nob Hill. Barely visible in the distance are ships on San Francisco Bay, Rincon Hill and China Basin. Along Market Street the familiar images of the domed Call Building, the steel frame of the Whittell Building, and, across the street, the Mututal Savings Bank (now called Citizens Savings Building) appear just to the left of the photograph's center. This scene demonstrates two important historical facts about the city: first, well before 1906 San Francisco was becoming a modern city with a skyline increasingly punctuated by massive steel-reinforced buildings; and second, the cranes visible at the sites of unfinished structures show that only months after the disaster, with rubble still piled along the streets, capital was already being invested in both restoration and new construction, which suggests great optimism about the city's rebirth. That local developers and workers barely missed a beat in the restoration and renewal of the city's badly decimated areas helps explain why San Francisco was largely rebuilt within three years of the earthquake and fire.

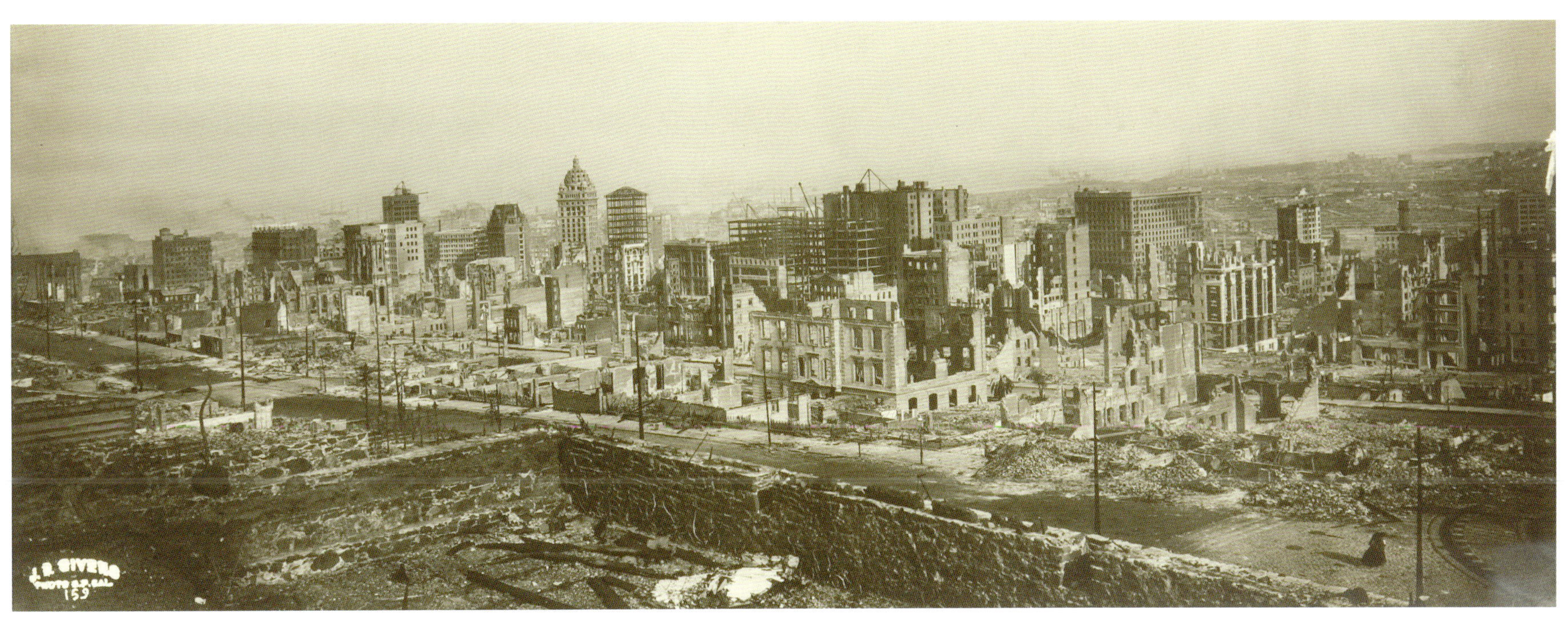

In this panoramic view of the western section of the downtown district, again taken from Nob Hill, Givens lets his camera sweep down in a southerly direction from the devastation in the foreground across Polk Gulch to the right, the Tenderloin, and Market Street, all the way to Potrero Hill and, in the far distance, San Bruno Mountain at the southernmost edge of the city. While Market Street from around Sixth Street west was not heavily built in 1906, the somber images of the devastated City Hall and Hall of Records are once more apparent here, looming like ghostly sentinels amid the fields of destruction. It was views like this that led the city's energetic and progessive ex-Mayor James Duval Phelan to assert that "San Francisco was no ancient city. It was the recent creation of the Pioneers and possessed the accumulated stores of only a couple generations. Its temples, monuments, and public buildings were not of conspicuous merit or of great value. There was, in fine, nothing destroyed that cannot be speedily rebuilt" (William Bronson, *The Earth Shook, The Sky Burned,* 158).

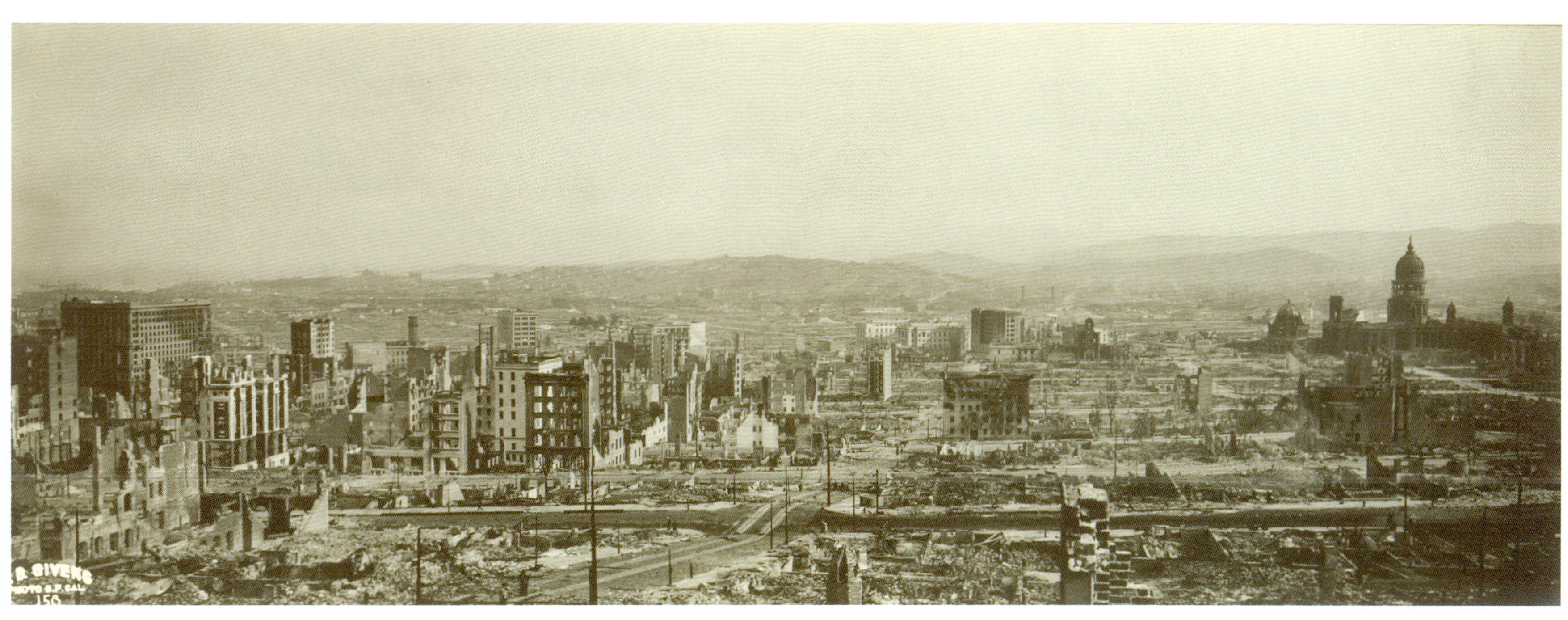

Using a panoramic camera pointing southwest from Telegraph Hill, Givens captures the crest of Nob Hill and the Call Building dome far below on Market Street. This photograph is one of several panoramic views he took of the downtown area, suggesting, perhaps, that he intended to combine the various panels to create a panorama of all the destroyed sections of the city. Givens would certainly have been familiar with earlier San Francisco panoramas dating from the 1850s by such photographers as George Robinson Fardon, Carleton E. Watkins, and Eadward Muybridge. From an elevated site on Telegraph Hill he provides a stunning view of the fire-destroyed residential areas on the crest and northern slopes of Nob Hill between Powell Street to the far left and Jones to the far right. The large building on the left is the Fairmont Hotel at California and Mason Streets designed by the Reid Brothers and near completion in April 1906. Though badly damaged in the fire, it was restored under the supervision of Julia Morgan and opened one year later. The smaller structure to the right is the mansion of silver millionaire James Flood that dates from 1886. It too was gutted by flames and later restored by Willis Polk in 1912, from which date it has been occupied by the Pacific Union Club. Both the Fairmont and the Flood Mansion survived the fire, unlike the great wooden mansions of Victorian nabobs around them, due to their masonry construction. Visible at the upper right of this view is the storage tank and ruins of the Union Gas Works that supplied fuel for the city's homes, buildings, and street lights.

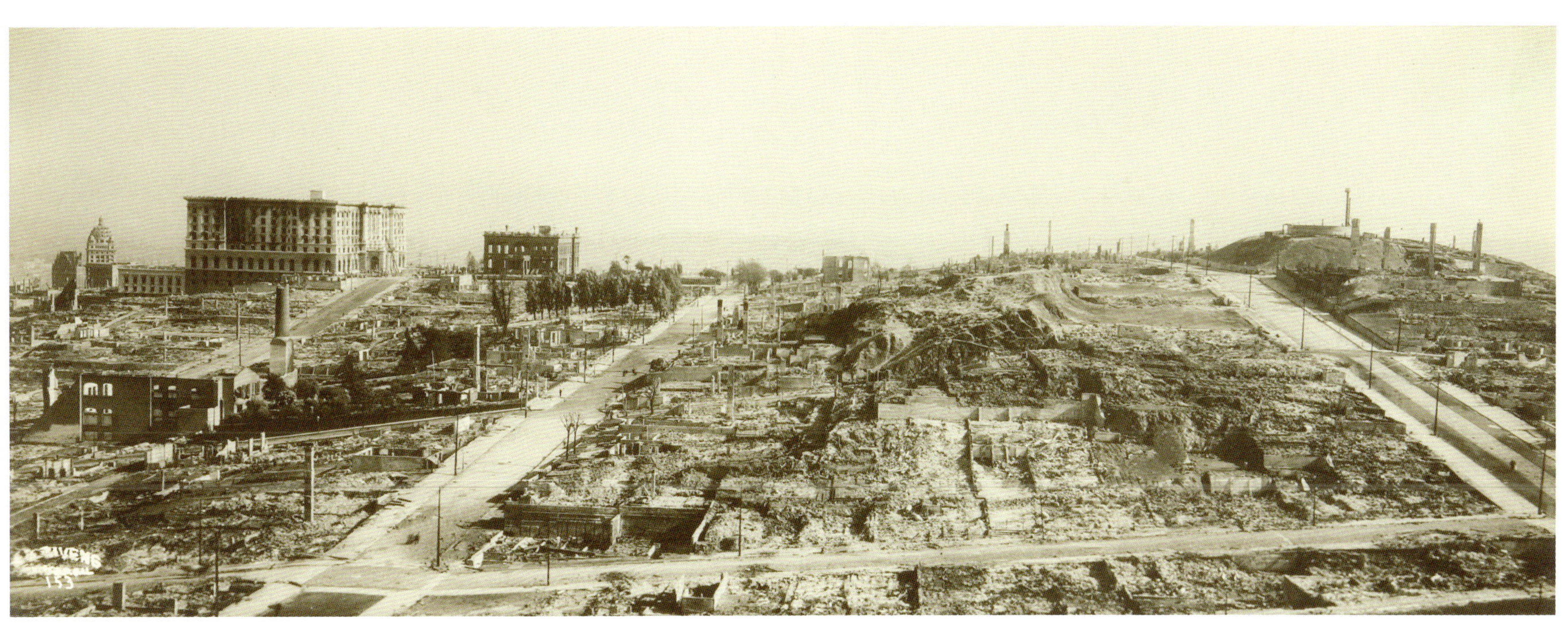

For this darkly desolate view Givens stands near Fifth and Mission Streets aiming his camera in a northwest direction in order to achieve an intimate sense of the destruction experienced in the heart of the city's densely built downtown area. Shattered walls, broken facades, piles of rubble, fractured cobble stone streets all attest to the predicament locals faced in restoring their community. In the emptiness of the foreground a mute and useless fire hydrant and an abandoned horse and carriage underscore the devastation left by quake and fire. In the background to the left stands the Flood Building at Powell and Market Streets, which was nearly completed at the time of the earthquake and withstood the tremor well due to its modern construction. At the upper right appears the decorative dome of the Call Building at Third and Market, which, though damaged by fire, stands at the same location today. The contrast between the vacuity of the foreground and the density of construction in the background creates a distinctive spatial aesthetic.

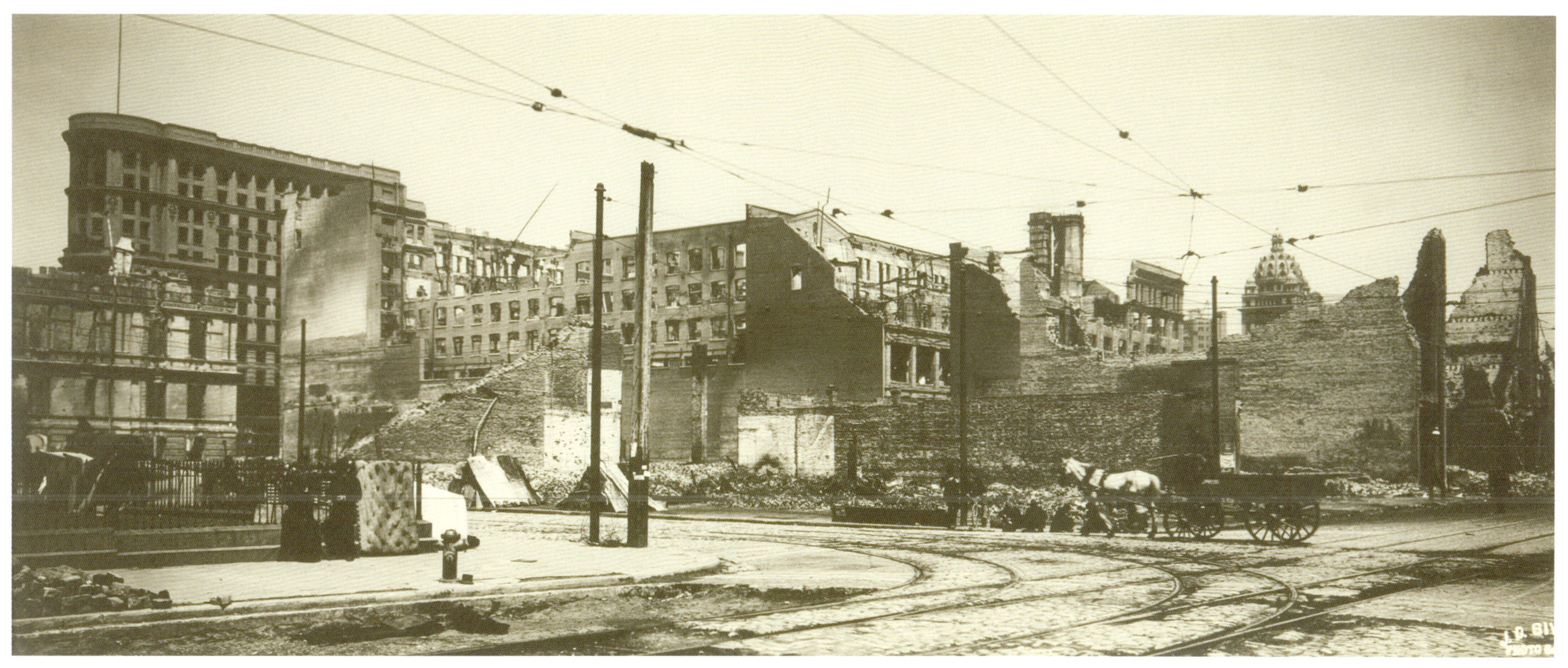

PLATE 21. James D. Givens, *Untitled,* 1906, gelatin silver print (Stephen Wirtz Gallery, San Francisco)

Though the original St. Francis Hotel dates back to the days of the gold rush, this magnificent second version photographed by Givens was designed by master architects Bliss and Faville and opened in 1904. During the two years before the earthquake, the new St. Francis quickly surpassed the three-decade-old Palace in its reputation as the most elegant hotel in the city. In this view we are looking west on Geary Street just east of Powell with Union Square in the right foreground. Givens illustrates with powerful imagery the combined destructive power of earthquake and fire. Geary is still strewn with tons of rubble, making passage by cable cars impossible. Buildings to the left have been decimated, and the St. Francis itself is charred from top to bottom. Like the Flood Building three blocks to the south, the St. Francis was built with a steel frame and was faced with gray Colusa sandstone. The original building had the two wings shown here adorned with rusticated walls and a decorative style consisting of Renaissance/Mannerist elements, such as two-story arches, Ionic columns, handsome balustrades, quoined corners, and an elegant neoclassical cornice rendered in copper. To the right of the building we see the steel truss work of a new third wing, which was begun before the earthquake and completed in 1907, about one year after this photograph was taken. Though the hotel was badly damaged by flames a day or so later, it continued normal operations after the tremor hit. Legend has it that in the chilly early morning air of April 18, Italian tenor Enrico Caruso made the five block walk to breakfast at the St. Francis, since his hotel, the Palace, was in a state of chaos following the temblor's first jolts.

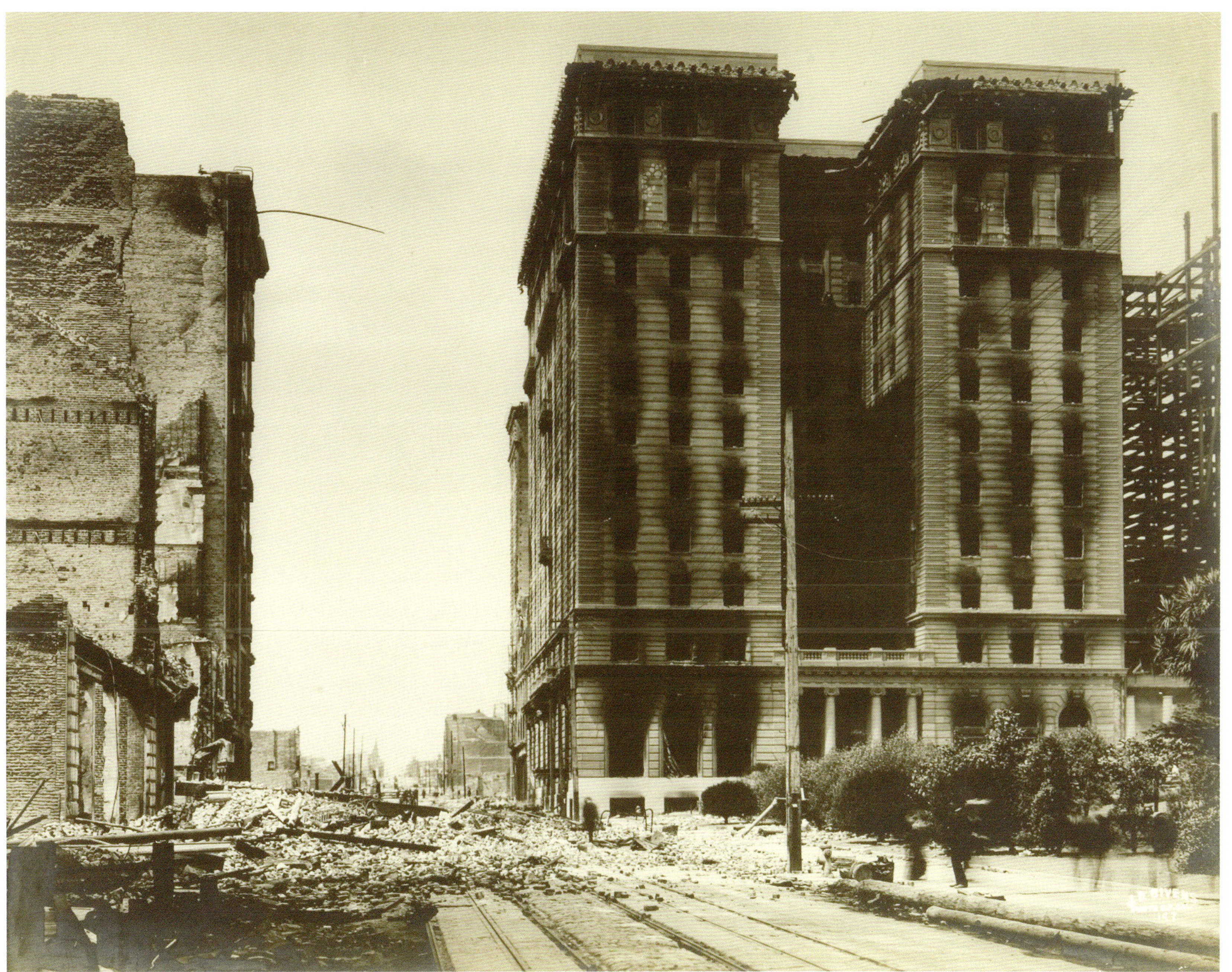

Givens stands near Sixth Street aiming his camera northeast along Market toward the Ferry Building in the distance. In this view, as in most others of the earthquake's immediate aftermath, we see San Francisco's streets still dominated by horse-drawn vehicles with nary an automobile in sight, though hundreds were present in the city, and they would prove their value in the wake of the disaster. The prominence of the horse in a place swiftly rebuilding itself in modern architectural styles points to the fact that even before the earthquake San Francisco was a community in transition from a Victorian city to a twentieth-century metropolis. Other Victorian remnants visible are the undisturbed cable car tracks and slot and the formality of dress worn by strolling pedestrians. In addition to the skeletons of commercial buildings in the foreground, Givens's lens also captures the Flood Building, still at the same location today on Market at Powell, and beyond that the Citizens Savings Building across from the Call Building three blocks in the distance. Note also the chair provided for those awaiting the arrival of trolleys.

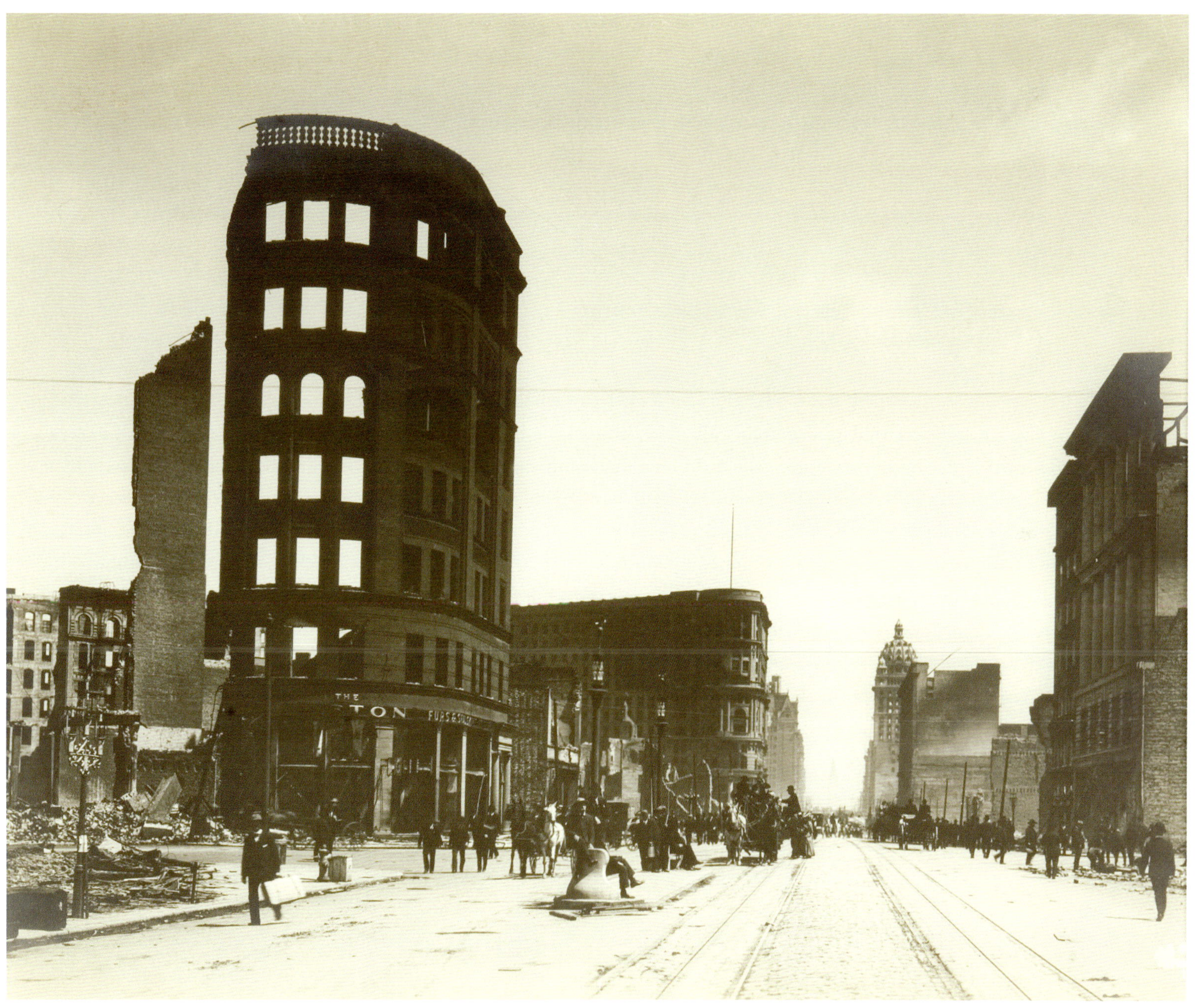
THE
TON FURSE...

The heavily clothed passersby indicate that it was a cool day when Givens, standing near First and Mission Streets, trained his camera across rubble, shattered walls, and fire-darkened, broken colonnades and arcades to capture one of San Francisco's first Chicago-style skyscrapers. In the center of this view stands the towering Mills Building, erected in 1891 by Bank of California cofounder Darius Mills and designed by Burnham and Root. Located at 220 Montgomery Street, two blocks north of Market, the Mills Building with its Romanesque façade of marble, brick, and terra cotta was regarded as one of the city's most important and imposing structures at the time of the earthquake. Though seriously damaged by the fire, its steel construction survived intact. In 1908 the architect Willis Polk, working with Burnham and Company, restored and enlarged the Mills. This was the first of several renovations he undertook over subsequent decades. Also pictured in this view at the far left is the fire-damaged but structurally sound Fairmont Hotel atop Nob Hill that was near completion in the spring of 1906. Photographs such as this one provide excellent evidence of the superiority of reinforced structures over unreinforced ones in surviving major earthquakes.

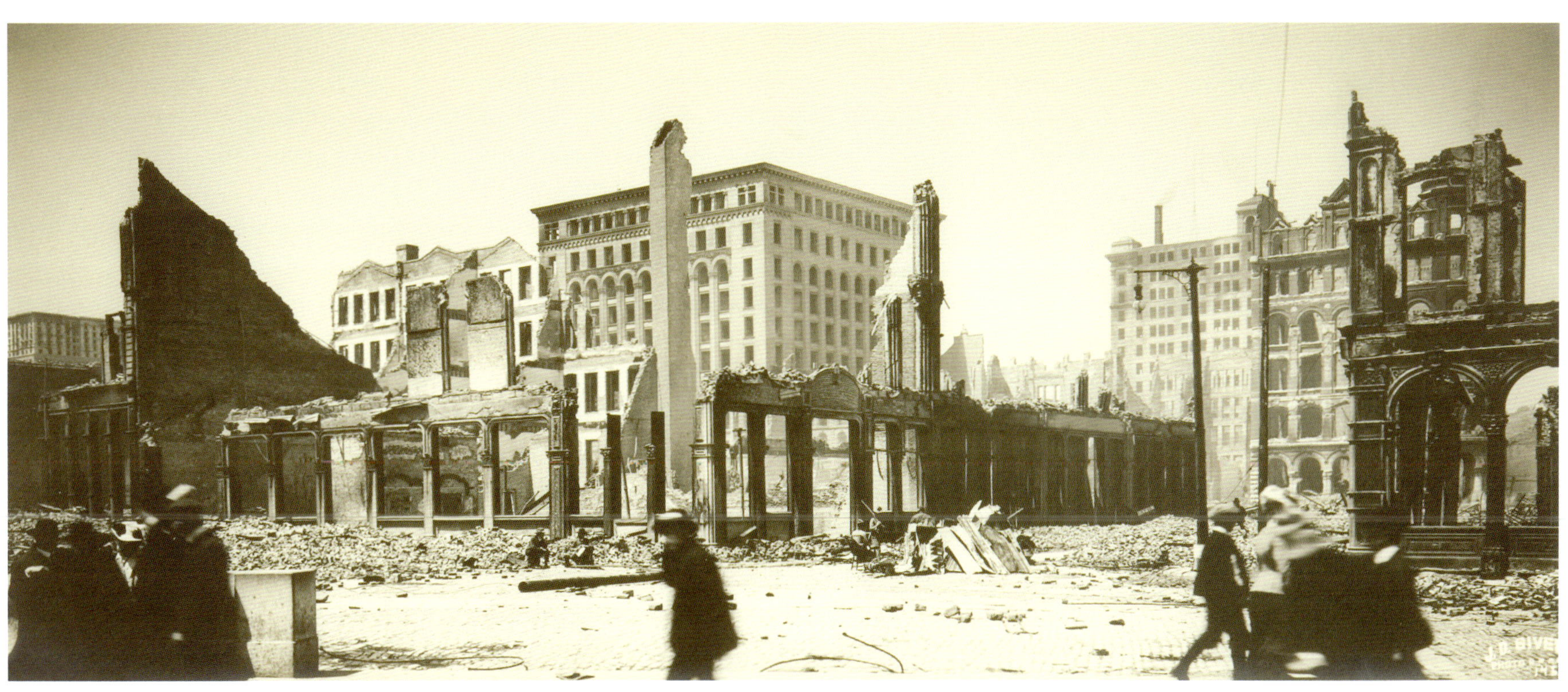

Edgar A. Cohen was a gifted amateur photographer who took the ferry across San Francisco Bay from his well-to-do family's Alameda home in order to make hundreds of images of the disaster stricken city. Though not a professional image maker, Cohen, a member of the Camera Club of California who had written articles for Camera Craft magazine, was undoubtedly moved by the visual possibilities of a great American city reduced to ruins with terrifying abruptness. For this view Cohen points his camera northeast from around Van Ness Avenue and Bay Street toward the slopes of Russian Hill. In the distant background to the right is a portion of the city's Market Street skyline and the ruins of City Hall and the Hall of Records. In the center stands a commanding, though badly damaged building framed by a row of Italianate Victorians to the left and a deeply shadowed neoclassical structure to the right. While the streets have been largely cleared, the piles of rubble visible in the scene indicate that Cohen's view probably dates from some weeks after the disaster.

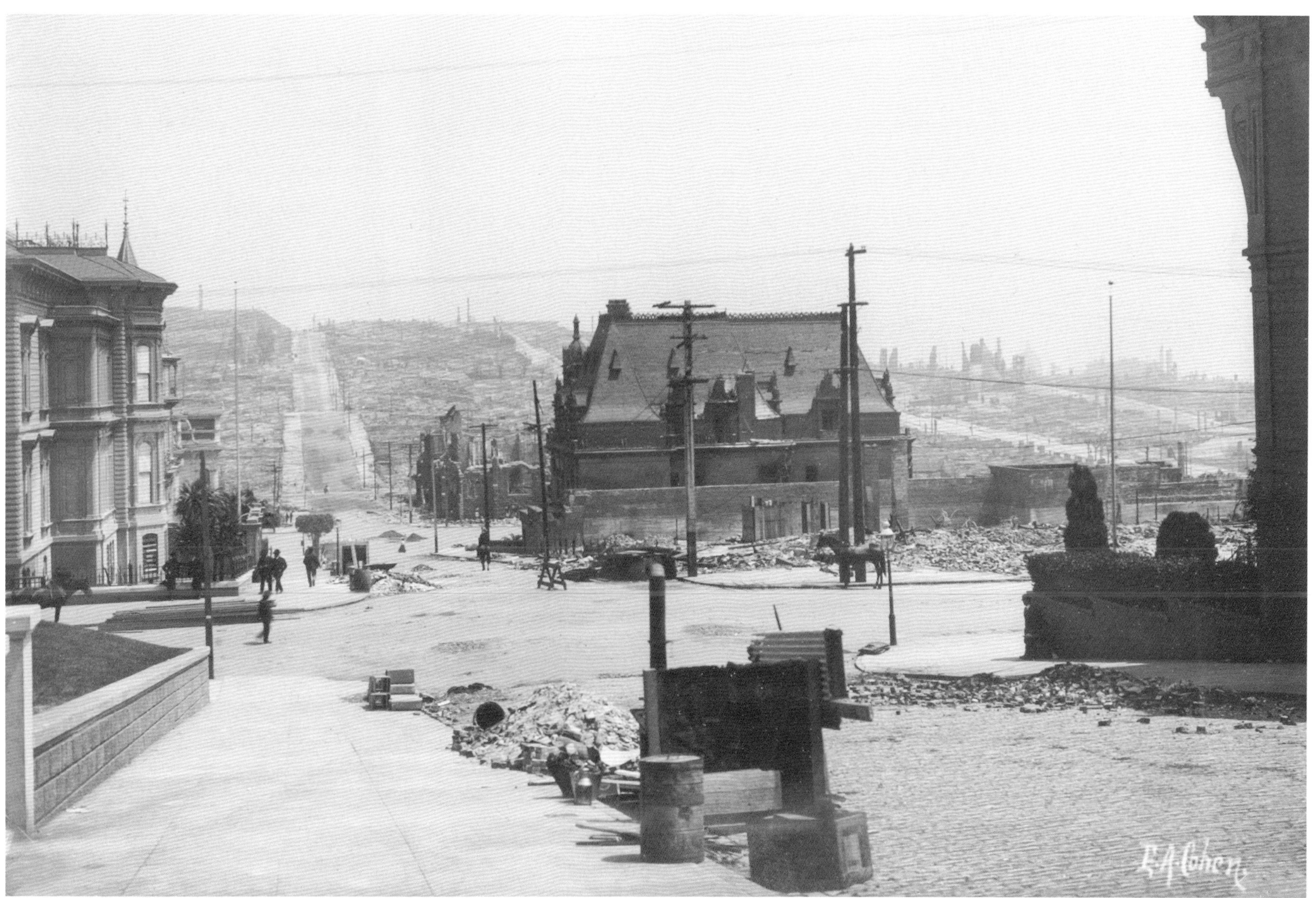
E.A.Cohen

Standing near the foot of Russian Hill, Edgar A. Cohen took this dramatic view of its southwestern slopes. Cohen uses the retaining walls both to lead the viewer's eye up to the crest of the elevation and to create compositional axes delineating the rubble fields ascending the precipitous hillside. In the distance, to the right, are the slopes of Telegraph Hill, and on the street in the middle foreground is the curious image of a well-dressed man bending over to pick something up off the pavement, perhaps included to give a sense of scale. Russian Hill, with its commanding views of the city and the Bay, had been the site of many homes from the gold rush on. In the late 1880s the Swedenborgian minister Joseph Worcester built the first of several shingle-style houses that were spared in the great fire. By the 1890s several of the houses visible in this view were occupied by local Bohemians, such as writer Gelett Burgess and architect Willis Polk. Cultural doyenne Kate Atkinson's house, which also stood atop Russian Hill, was a center for get-togethers of the city's artistic and literary elite in the decade and a half before the great disaster. The large shingle structure standing on the peak of the hill is the home of water company and transportation mogul Horatio P. Livermore.

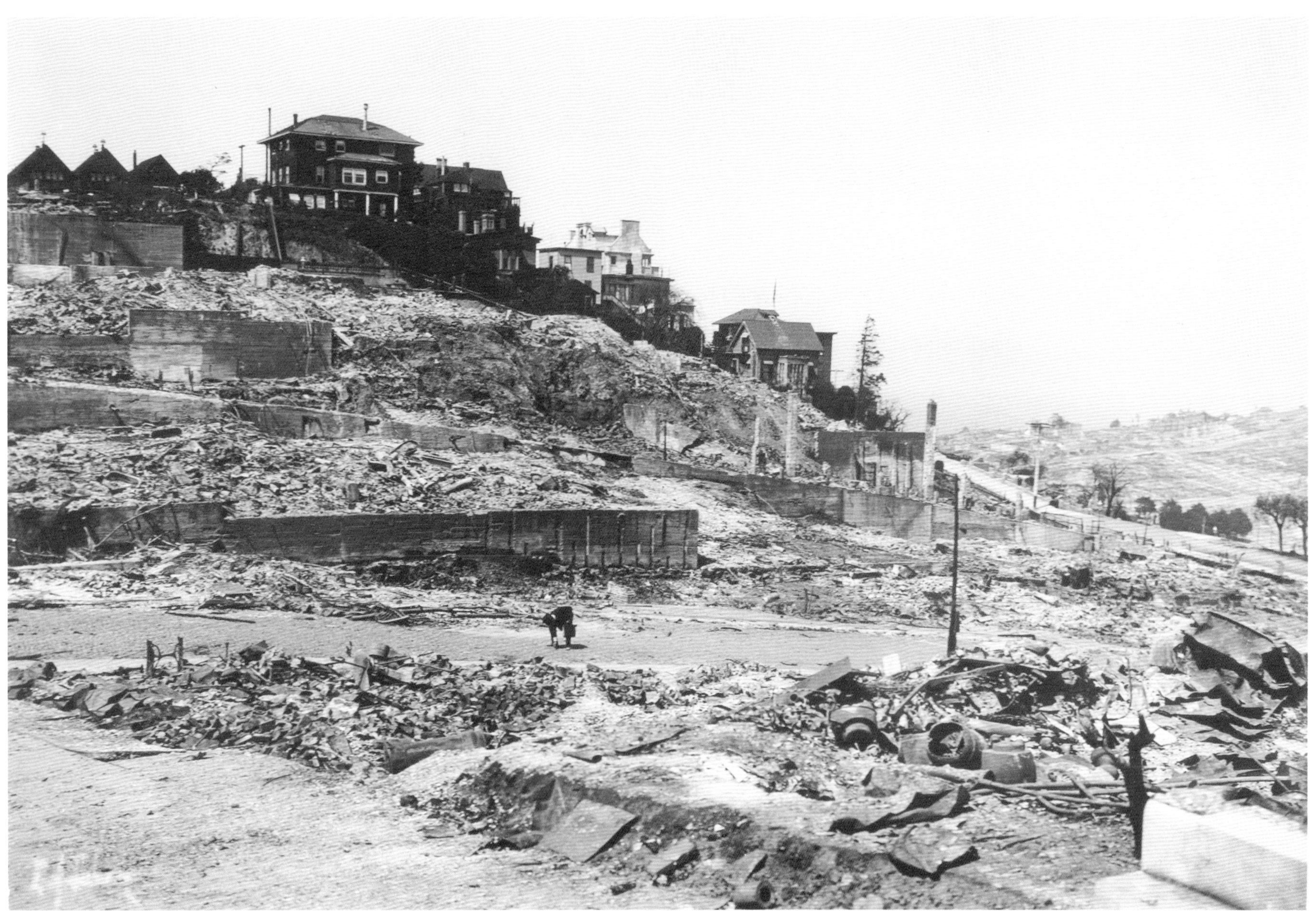

Arnold Genthe published this highly atmospheric photograph with the title "Steps that Lead to Nowhere." The spectral sense of a vanishing past is conveyed by the disembodied stairway that stands atop a desolate site on Nob Hill. This image is a fine example of how pictorialist techniques of composition and exposure can create a powerful visual statement of mood. Genthe's lens dwells on destruction, the ruined mansion in the foreground and the City Hall tower in the distance. The soft focus of the camera suggests an almost viscous atmosphere evocative of desolation and impermanence. Though somewhat theatrical in conception, this view is convincing evidence that a skilled photographer has remarkable artistic latitude in representing the material realities in which humanity exists.

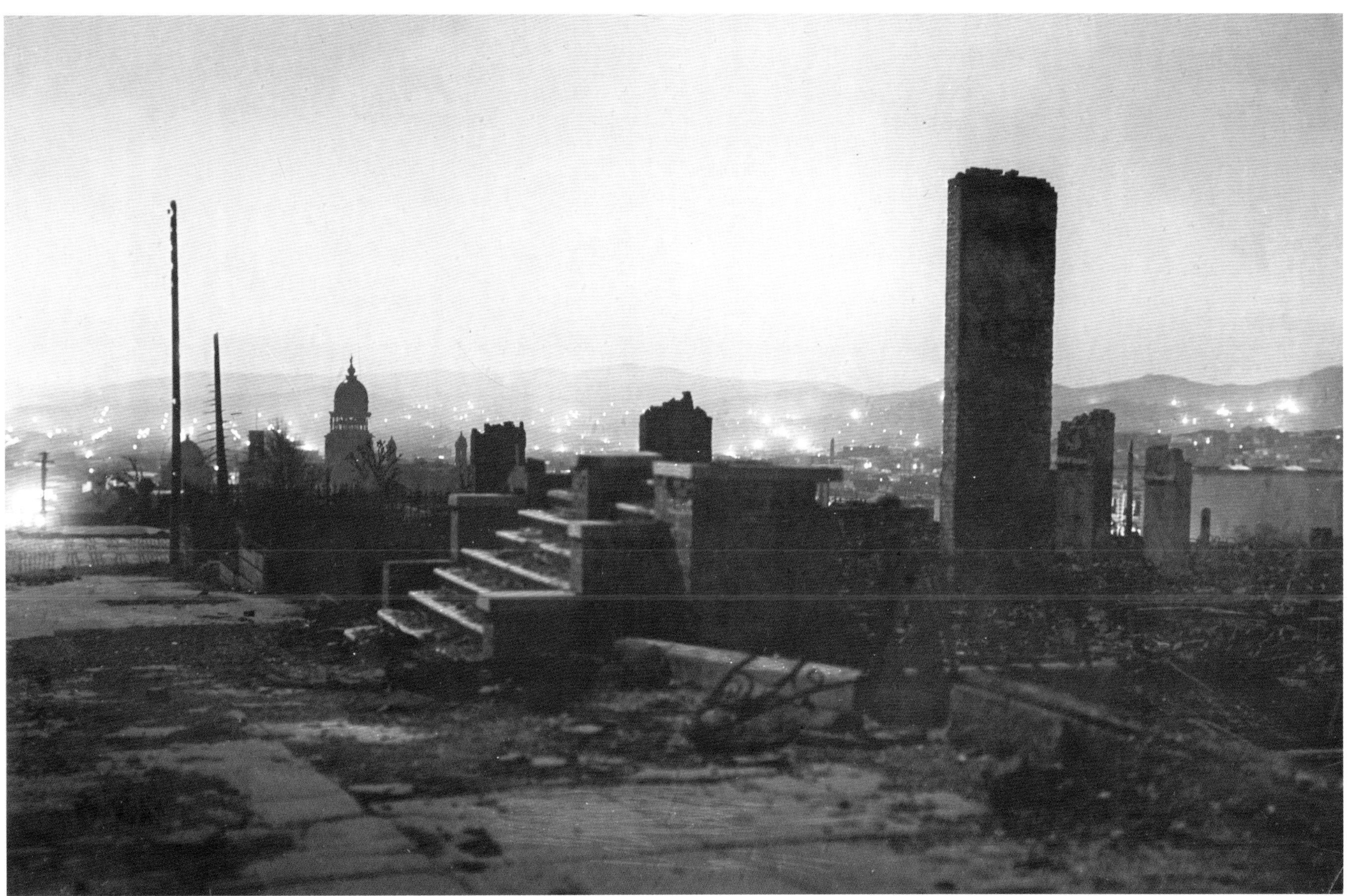

Willard Worden's often reproduced photograph, "Portals of the Past," is probably the single most powerful evocation of nostalgia for the lost San Francisco. The impressive marble veneered Ionic portico had served as the entryway of the now destroyed A. N. Towne mansion on Nob Hill. Its elegance stood in stark contrast to the ruins that lay everywhere around it, ruins created by the destruction of many of San Francisco's most palatial mansions. The portal itself reflects the profound sense of emptiness that San Franciscans had experienced in the wake of the earthquake and fire. But Worden uses composition to transcend simple nostalgia for "the city that was" by casting this small architectural fragment as a graceful classical frame for the devastated City Hall tower in the background. The combination of public and private loss represented in this image suggests the breadth of the challenge that was faced by San Franciscans in the rebuilding of their city. The portico itself would come to function as a powerful symbol for San Francisco. In 1909 a citizens' group, led by ex-Mayor James D. Phelan, moved this sole remaining remnant of the Towne mansion to the shore of Lloyd Lake in Golden Gate Park.

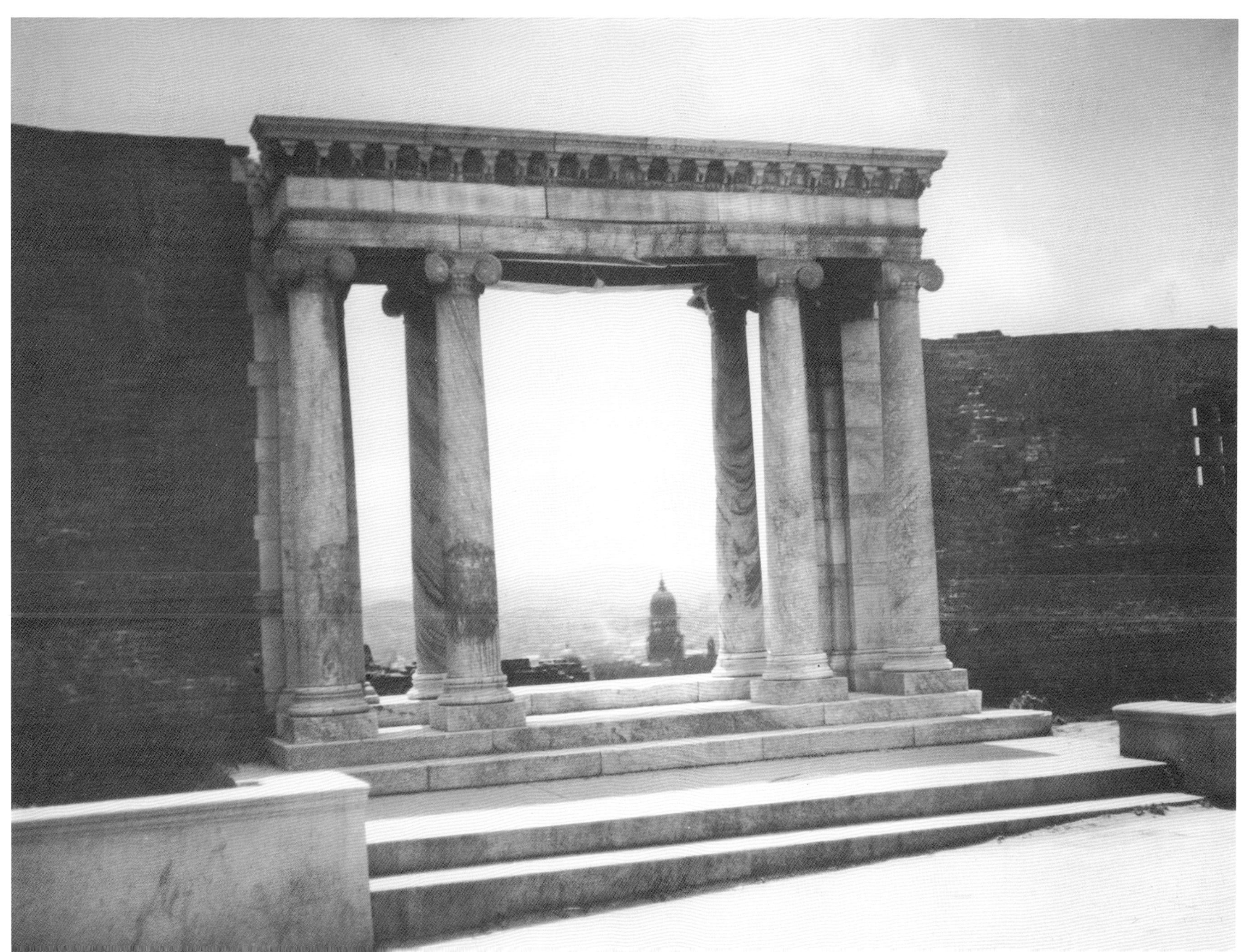

In this view Army Signal Corps photographer Henry Chadwick demonstrates not only the devastation in the wake of the disaster but also the quotidian activities that thousands of citizens were faced with in the recovery effort. Here the citizens are members of Chadwick's family. The long shadow of the figure with the pail suggests that this photograph was taken late in the afternoon some days after the fire. The casual postures and relaxed facial expressions no longer signify fear, melancholy, or befuddlement, but rather an ordinary family chore in the new circumstances brought on by the recent catastrophe. The untucked blouse of the pregnant woman, the simple hairdos, work shirts, and rolled up sleeves typify a brief period in local history when middle and working class San Franciscans had to make do living and laboring in the out-of-doors. The buckets and kettles shown in this scene are perhaps related to the task of rebuilding the brick foundation for a new home to replace the one that was destroyed.

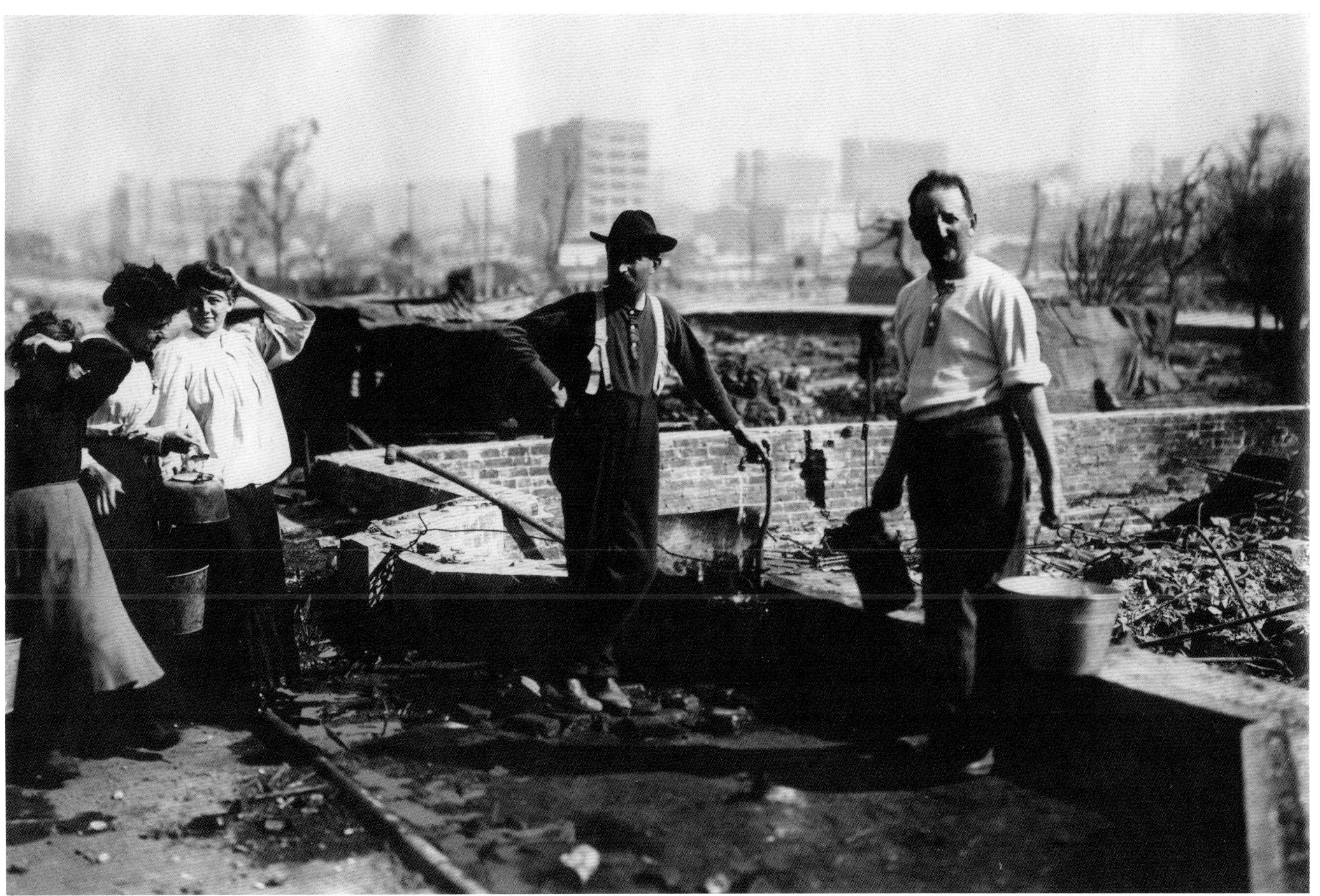

Here, Chadwick uses clever compositional elements both to frame the figures in this photograph and to illustrate their isolation from the mass of urban humanity. Atop some city hill, perhaps Russian Hill, a mother and her two daughters swing and stand within the frame of a play structure. The ghostly image of the Fairmont Hotel on Nob Hill is visible in the left background, but a long metal fence is seen on the right and a small, partly shadowed tent creates a sense of separateness and the necessity of self-reliance at a time of crisis. The pathos of the image is not so much attributable to the psychology of the three figures in the photograph's middle ground as to the general situation in which they find themselves.

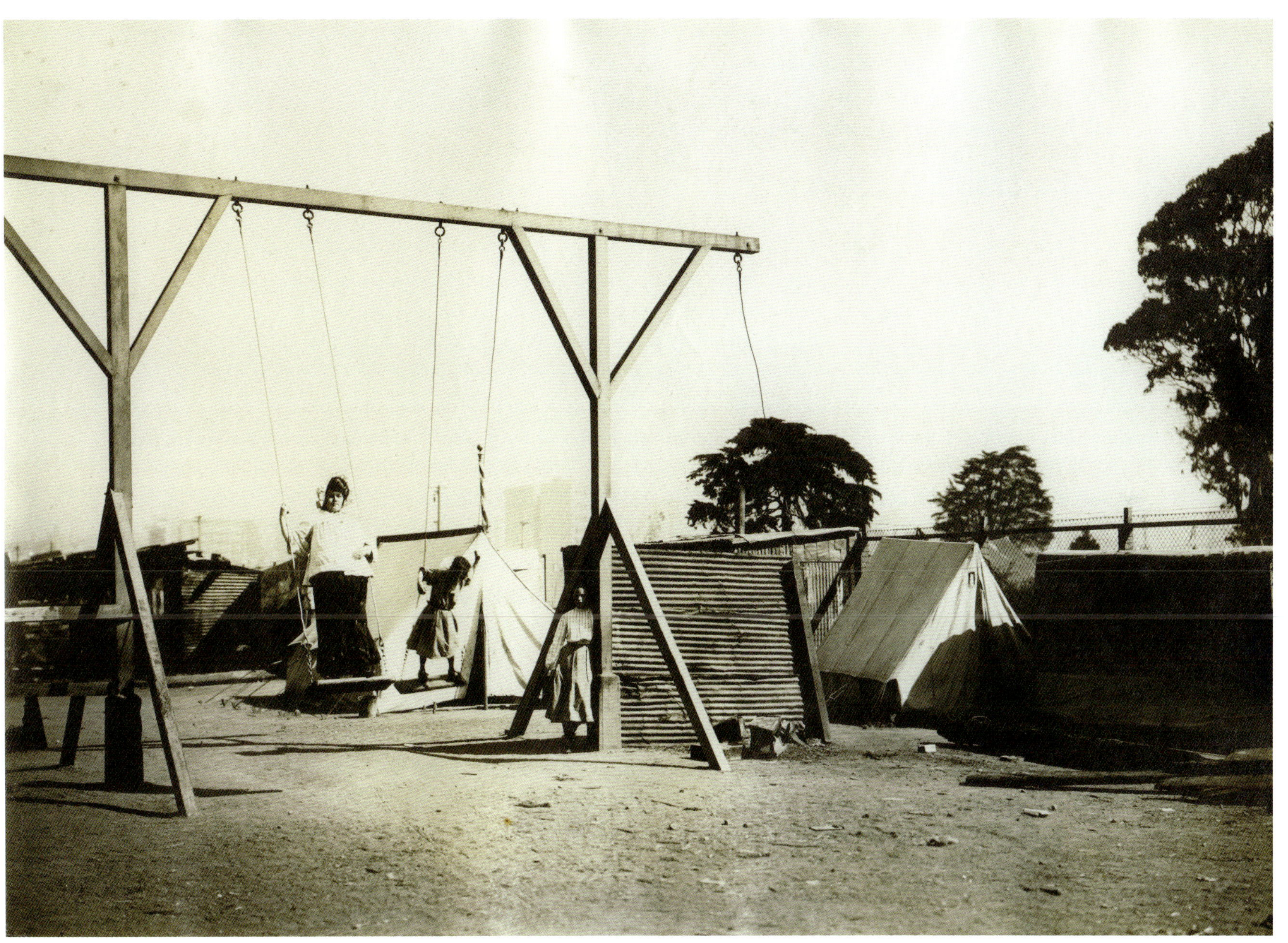

PLATE 30. Henry W. Chadwick, *Untitled,* n.d., gelatin silver print
(Visual Studies Workshop, Rochester, NY)

In this photograph, Henry Chadwick captures something of the determination of the newly homeless San Franciscans to persevere in the face of disaster and loss. Making ramshackle shelters out of pieces of timber and corrugated metal, the variously clad refugees pose for the camera with a combination of smiles and earnest expressions. Although an unprepossessing site on which to settle, the sandy, scrub-covered terrain offers safety from the dangers present in the central part of San Francisco. While clearly recording this moment to show all phases of life present in the crippled city for government purposes, Chadwick subtly directs the viewer's eye to the flimsy huts in the middle ground by including a semicircle of larger buildings surrounding them. Thus, in an otherwise inchoate pictorial field, he brings some semblance of structured composition.

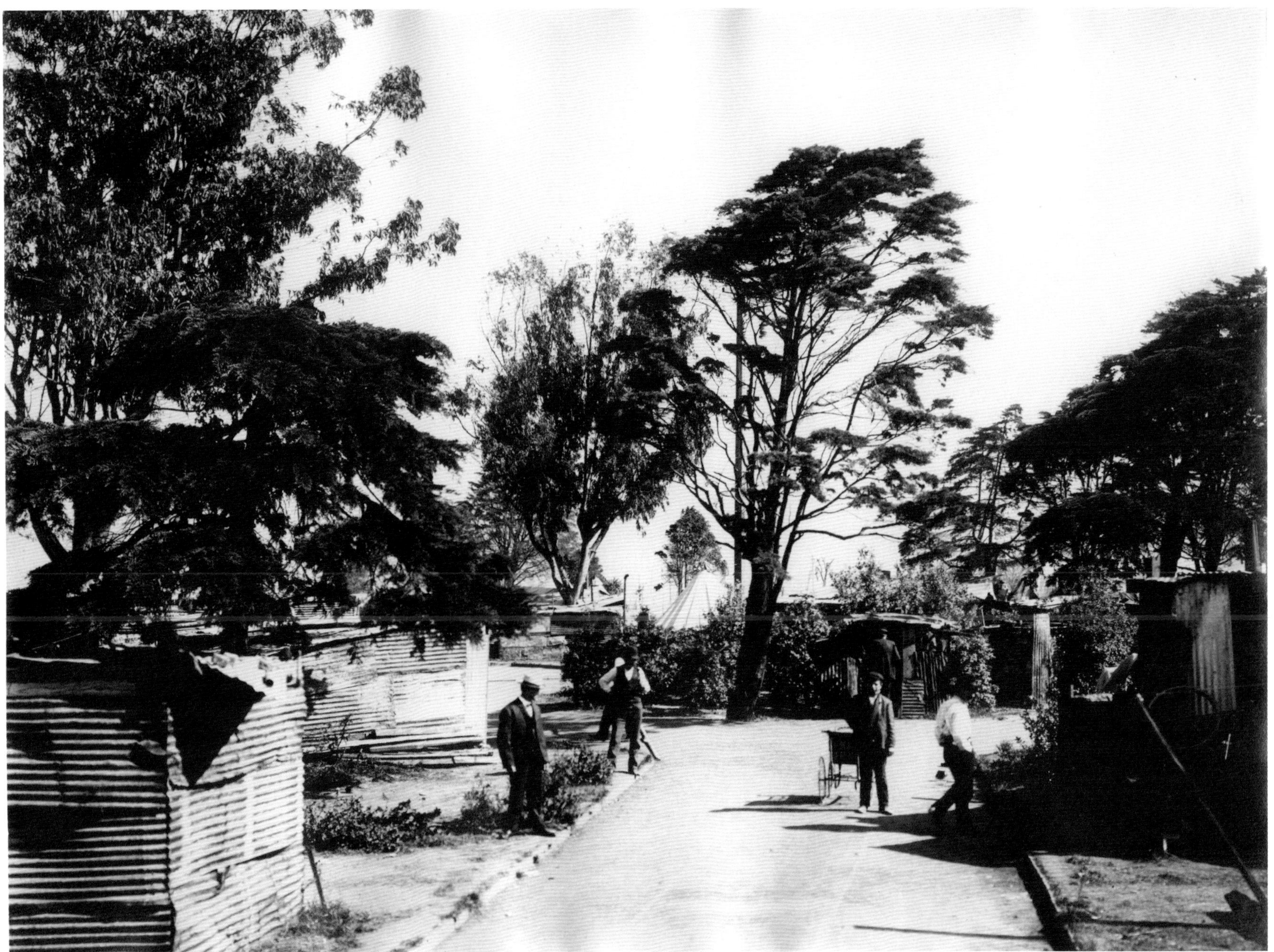

In the weeks after the disaster, before formal refugee camps began to take shape, San Franciscans who lost their houses and rented rooms had to scramble for lodging. Adequate shelter was a particularly urgent problem for large families. Here, a group of thirteen people, perhaps comprising one extended family or two neighboring families, pose in their new dwellings, two hastily erected lean-tos that are still under construction as the photograph was taken. Amazingly, even at this dire period in the city's life, several of the figures pictured are smiling broadly for the camera. The photographer stands several yards back from his subjects to include the geometry of hillside structures, paths, and patches of scrub surrounding their make-shift dwelling. A sense of isolated self-reliance sings forth from scenes such as this one.

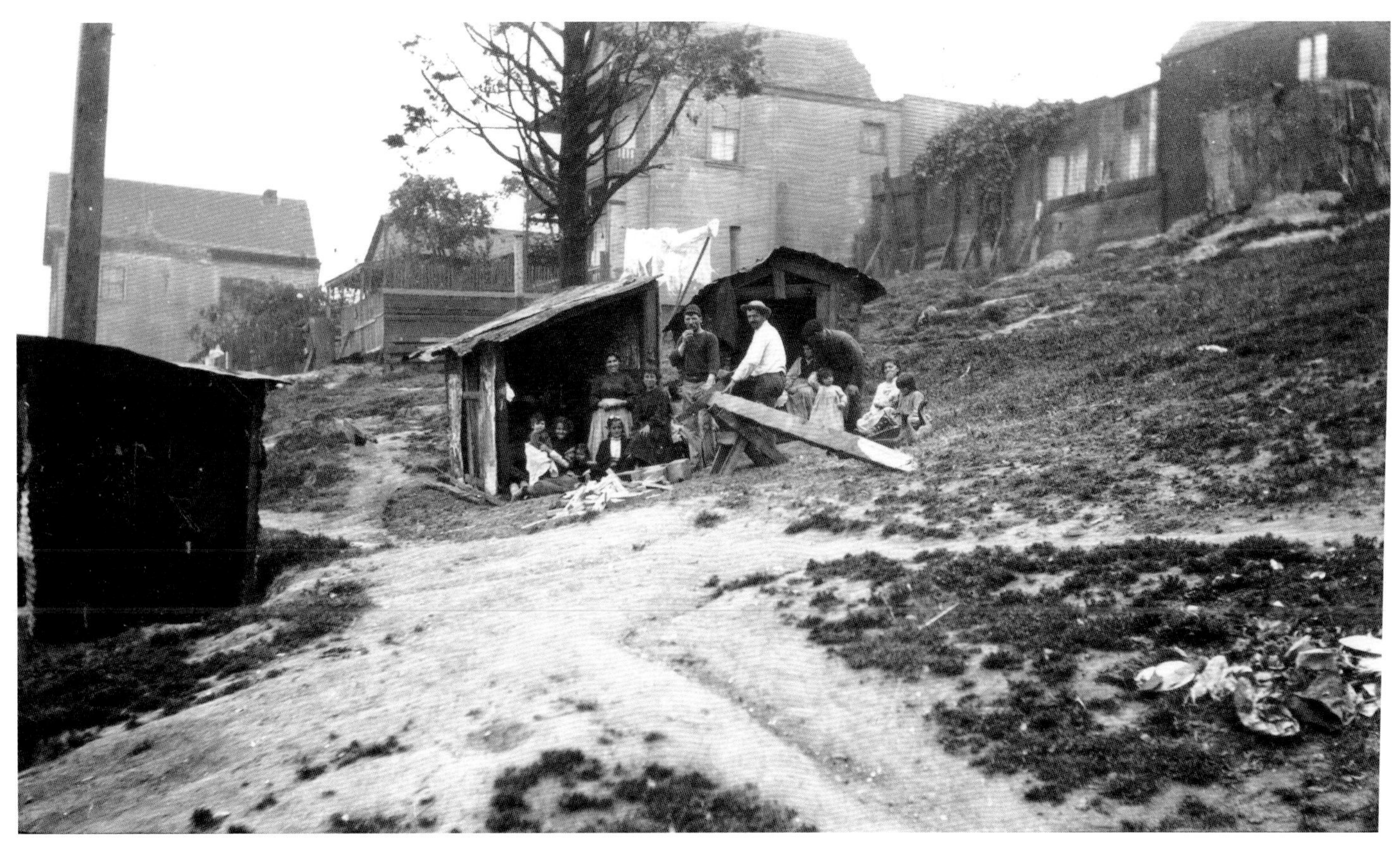

Images of Hiroshima and Munich at the end of World War II are evoked by this Chadwick view of a solitary Presidio soldier who stands watch while leaning against the remains of a lamppost amid blocks of rubble and ruins. The dangers of building with unreinforced brick in earthquake country are gravely apparent here. The photographer stands on Sutter Street west of Van Ness Avenue looking northeast. The Union Gas Works storage facilities are in the middle left background and Nob Hill, with the Fairmont Hotel perched atop, is in the middle right background.

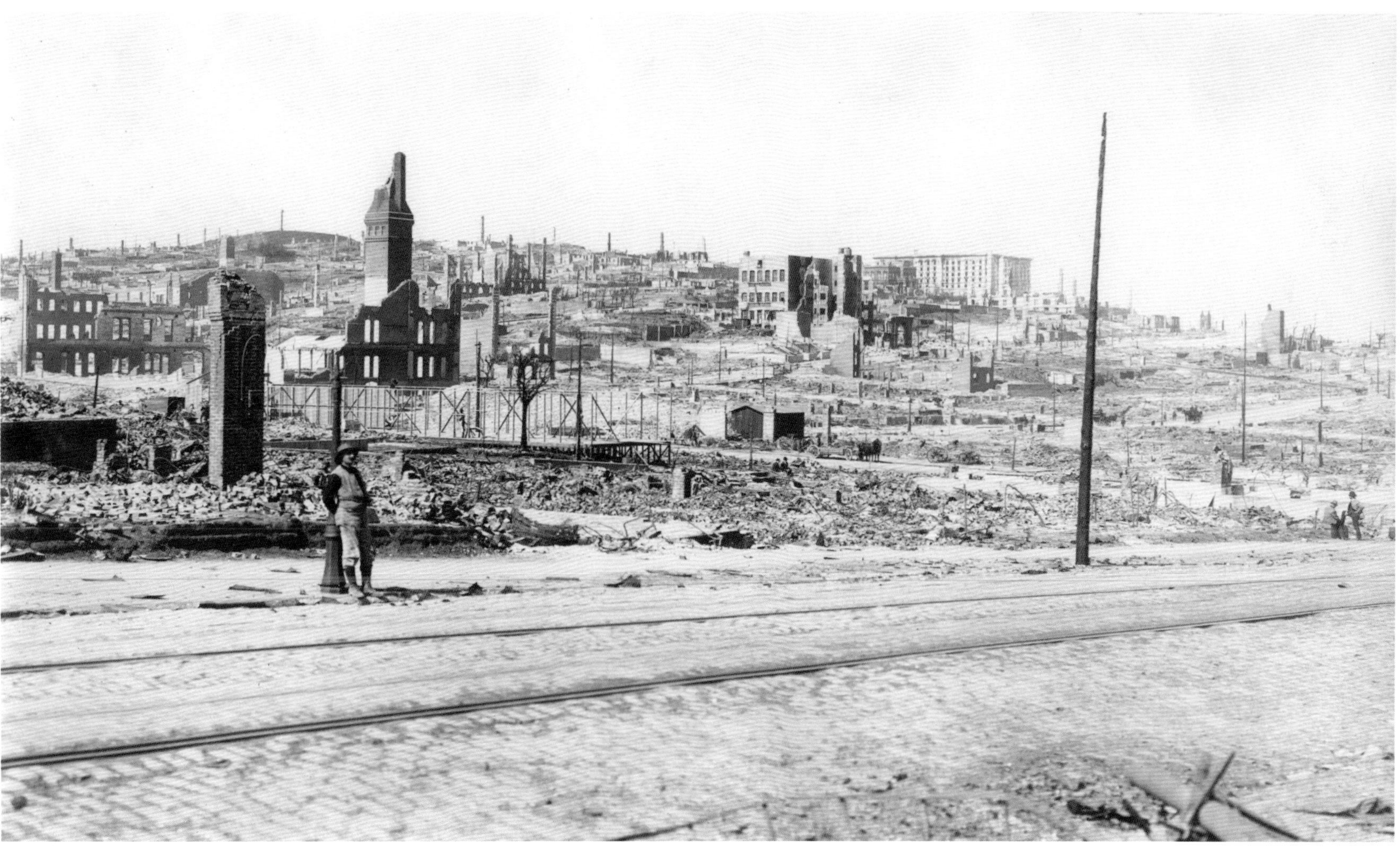

Chadwick took this view of Sacramento Street at Mason Street looking west from an upper floor of the unfinished Fairmont Hotel on Nob Hill. The transformation of this neighborhood, home to San Francisco's wealthiest silver and railroad barons, is apparent. In the left foreground is the rear entrance of the Connecticut brownstone Flood mansion dating from 1886. Though saved from destruction by its stone walls, it was gutted by flames and would later be purchased from James Flood's family by the exclusive Pacific Union Club. In 1912, architect and club member Willis Polk would renovate and expand the building. On the next block is a tent settlement lying within the foundation of the now vanished Collis P. Huntington mansion. Across Sacramento Street the camera captures blocks of rubble with only the occasional chimney to mark where great houses once commanded stunning views. In the upper right is the storage facility of the Union Gas Works. Nob Hill sits on fairly solid stone deposits, and this view shows that the cable car tracks on Sacramento have not been deformed by earth movement. Viewing the newly flattened terrain of his city, University of California President Benjamin Ide Wheeler, who saw the hand of divine providence in the destruction of San Francisco, came to this insight which he reported to the *San Francisco Chronicle* eleven days after the great quake: "I never appreciated so much the magnificent situation of San Francisco as today when I walked over it and saw the rectangular slashes of streets obliterated by the removal of disfiguring buildings."

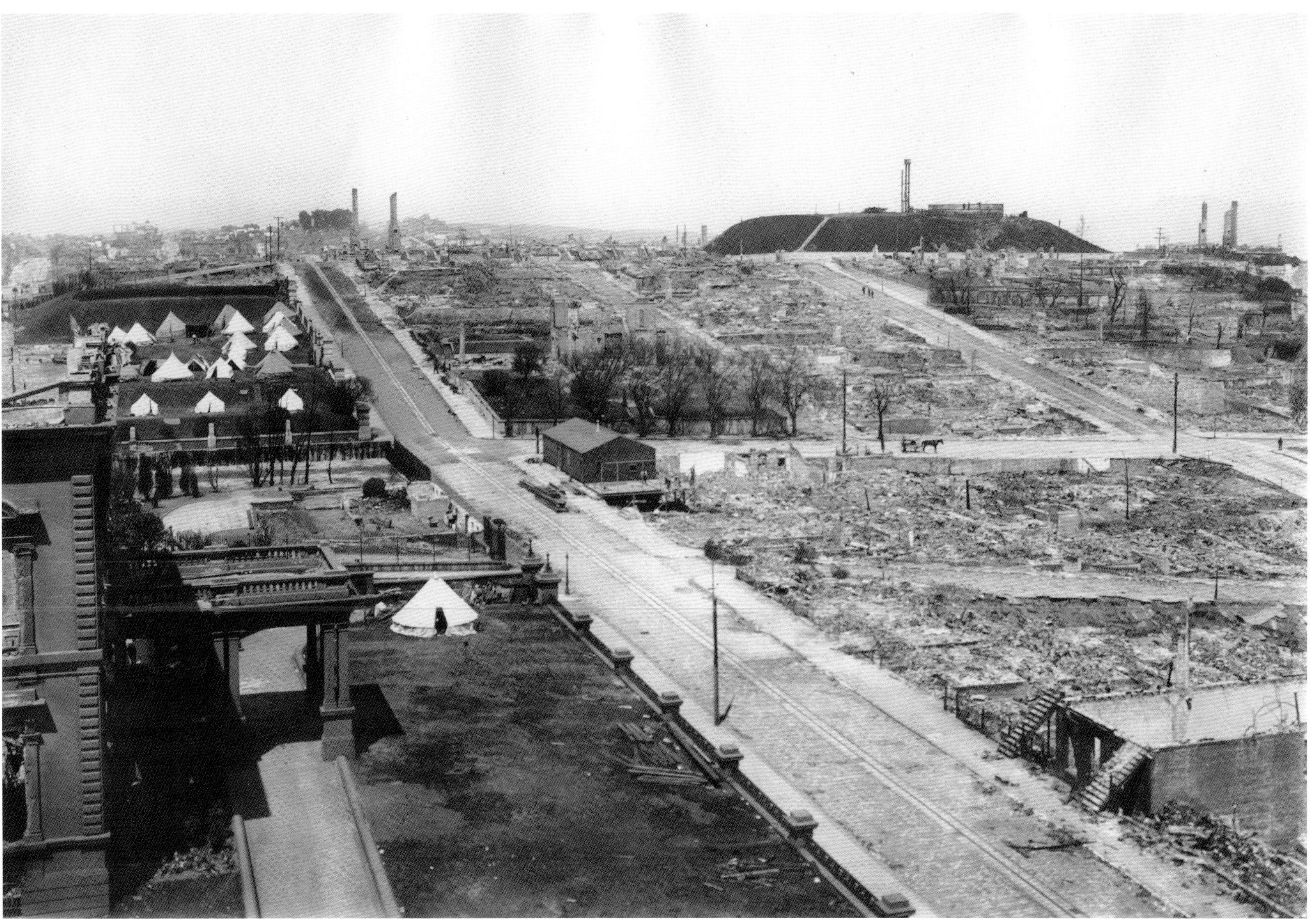

Some weeks after the disastrous events of April, Chadwick's camera reveals a bit of whimsy returning to the battered city. Here, in an area bordering downtown on the northwest, a slim young man stands delicately balanced atop the ruins of a fireplace. The playfulness of the pose is accented by the relationship of the man's outstretched arms to the limbs of the three badly burned trees to his left. Brick rubble has been moved to the side of the street, a small government-issued prefabricated "refugee" house stands where a large brick structure once stood, and the exterior of a damaged church looms behind.

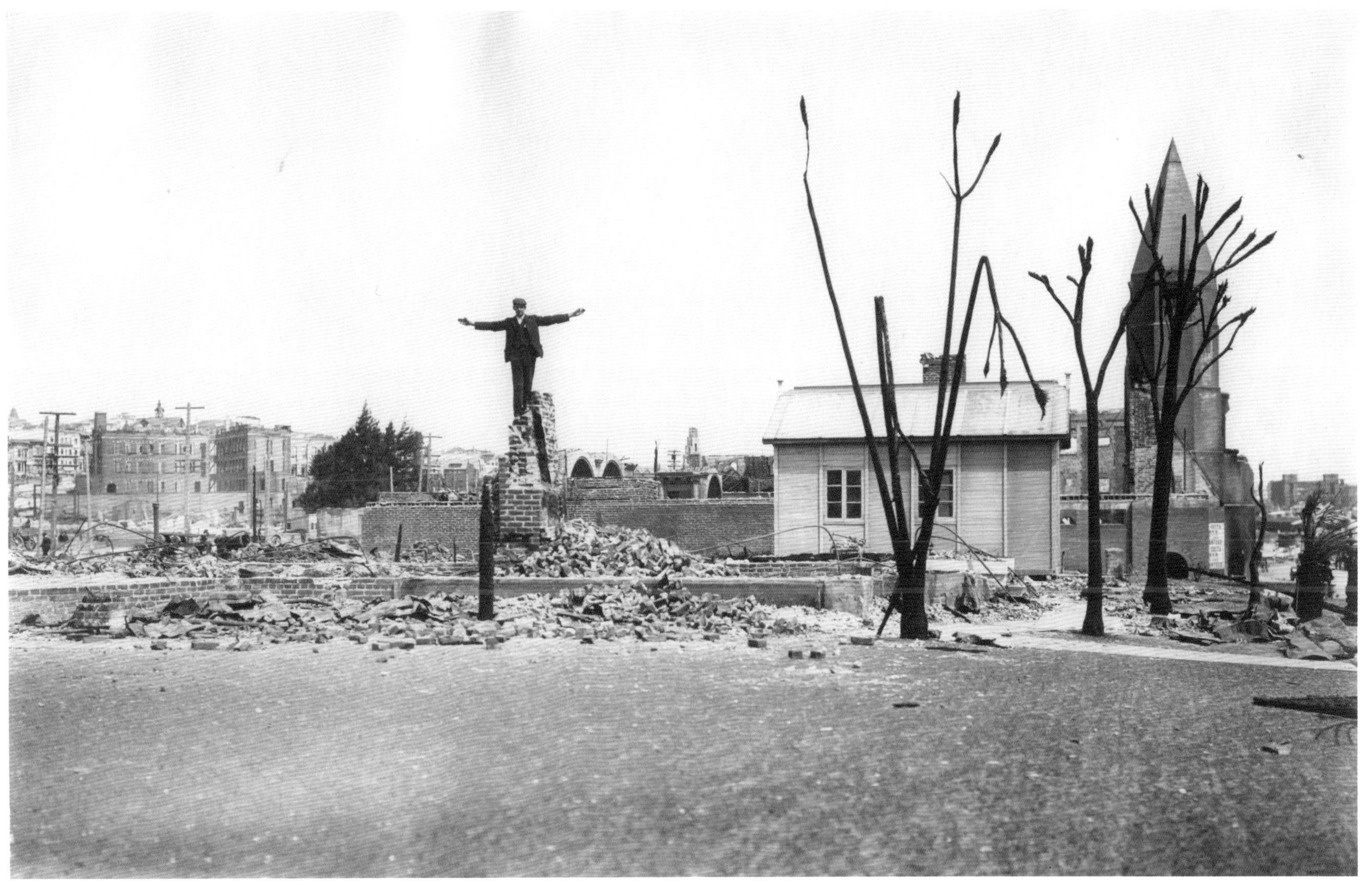

In the following series of five photographs taken some time during May and June of 1906 and documenting the United States Army's relief efforts, cameraman James Givens illustrates the stages by which food supplies were distributed to the local citizenry. Givens, who had already taken many pictures showing the immediate consequences of the disaster both to people and buildings, here turned to the task of demonstrating the efficiency and timeliness of the military relief effort and recording these views for official Army use. Though a civilian photographer, Givens maintained a studio at the Presidio because of the frequent commissions for work he received from local military officials.

PLATE 35. James D. Givens, *Issuing Rations Relief Depot No. 1 The Presidio,* 1906, gelatin silver print
(National Archives)

The issuing of rations is the subject of the first photograph. We view the process of loading teamster wagons within a panoramic picture field stretching from the dirt road in the foreground along the length of the Relief Depot building to the Bay and Marin Headlands in the distance. Here, Givens chronicles the cooperative efforts of military personnel and contracted civilian workers in the distribution effort. Interestingly, no motorized vehicles appear to be involved in the activity.

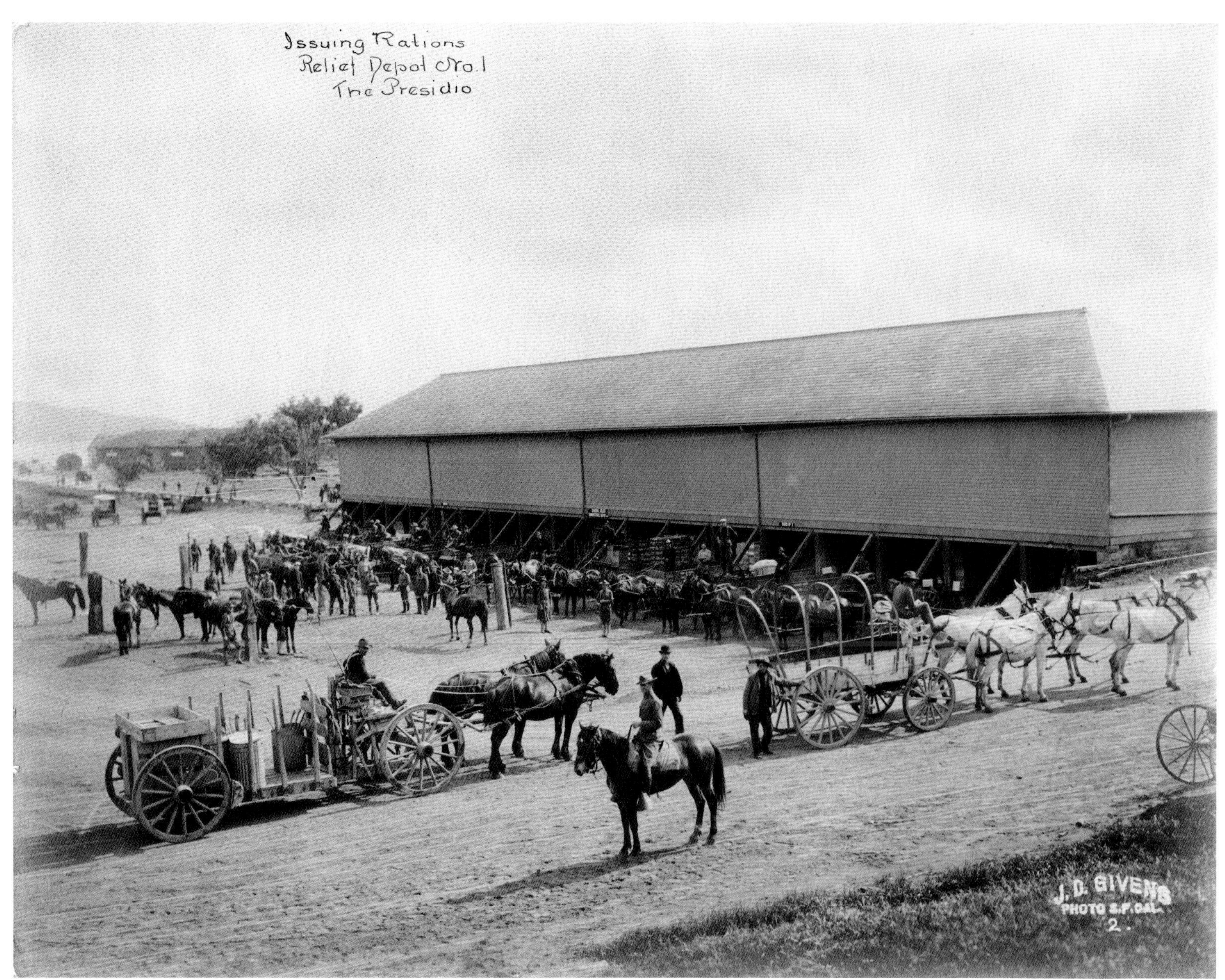

Issuing Rations
Relief Depot No. 1
The Presidio
J.D. GIVENS
PHOTO S.F. CAL.
2.

PLATE 36. James D. Givens, *Potato Warehouse Relief Depot No. 1 The Presidio,* 1906, gelatin silver print
(National Archives)

The second photograph shows hundreds of bags stored in a potato warehouse where they will eventually be loaded onto wagons and transported to various distribution points in the city. To the far right we see an armed guard providing security, large cloth tarpaulins protecting other foodstuffs, and a portion of a building housing Presidio offices. The pattern seen in the six building openings and stacked bags undoubtedly caught the photographer's attention as an element of abstract geometric design within an otherwise ordinary human activity.

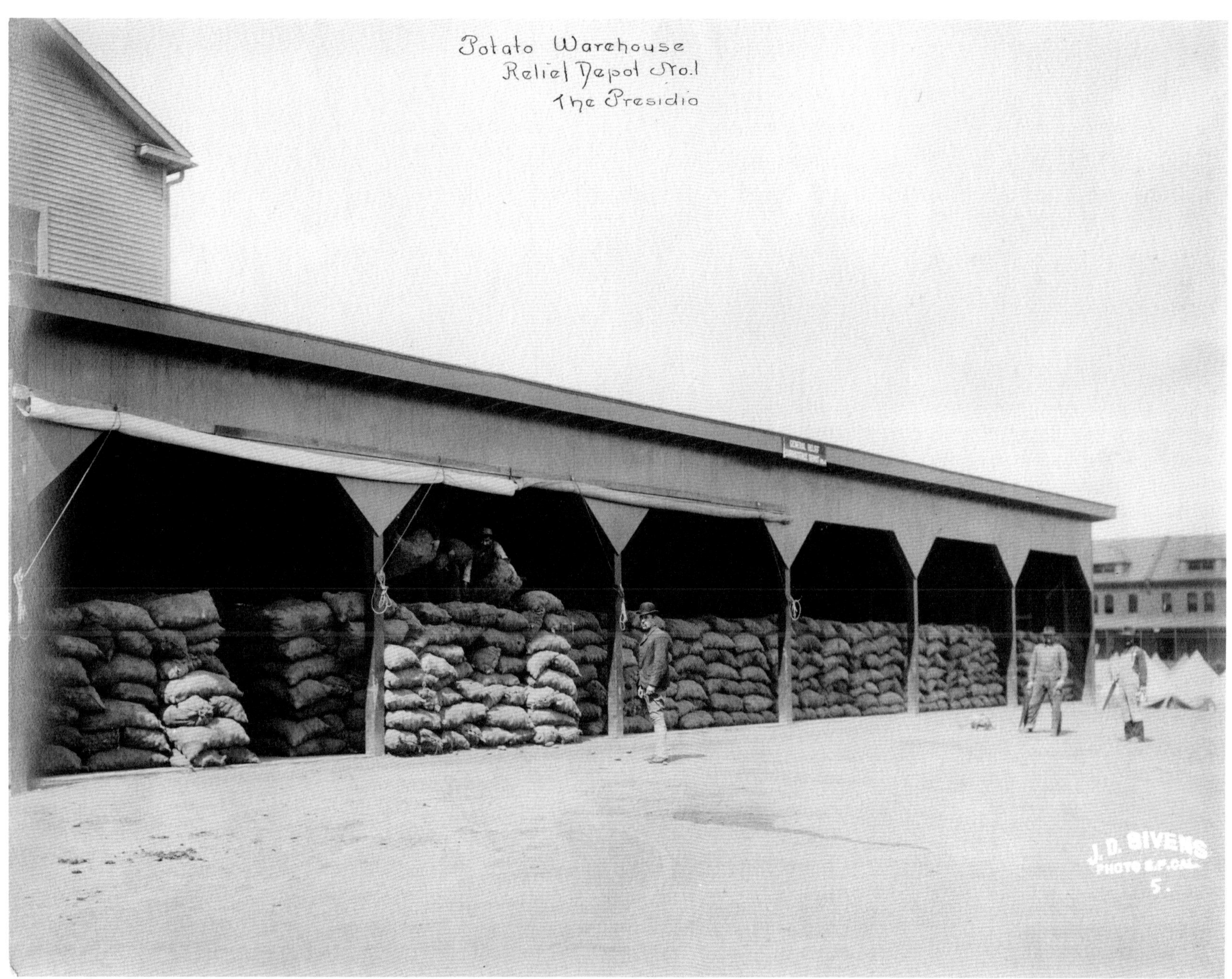
Potato Warehouse
Relief Depot No.1
The Presidio

In the third photograph we see a group of military men, civilian workers, and clerks standing in front of the General Relief Subsistence Depot at the Presidio. This massive clapboard warehouse was used for storage of food and other supplies shipped to San Francisco from elsewhere in the state and nation. While it is clear that the eighteen men in the scene know that they are being photographed, their casual postures imply that they are engaged in now routine daily tasks rather than the emergency efforts of the first days after the disaster.

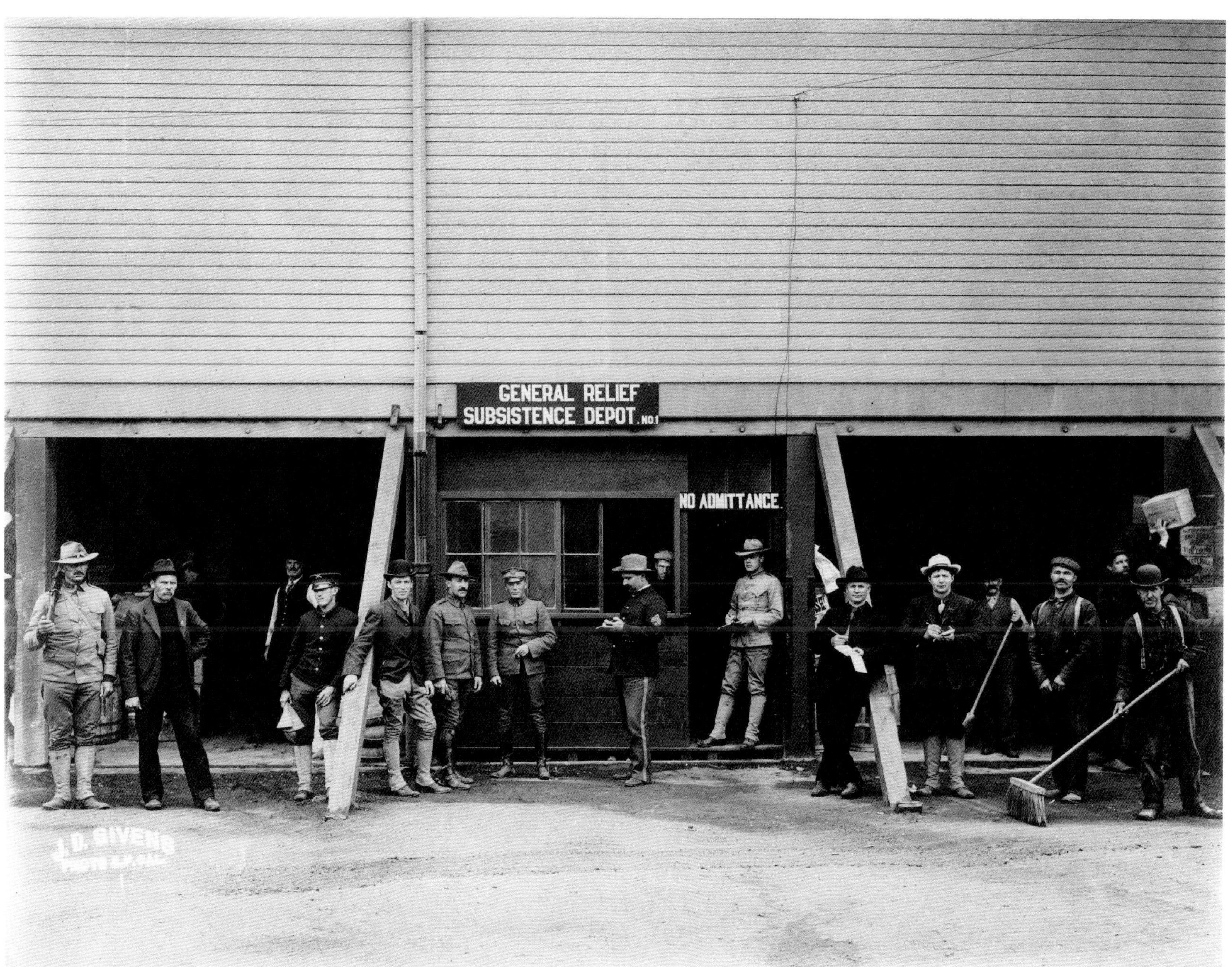

GENERAL RELIEF
SUBSISTENCE DEPOT. NO.1
NO ADMITTANCE.
J.D. GIVENS

PLATE 38. James D. Givens, *Stacking Flour Moulder Warehouse Relief Depot No. 3,* 1906, gelatin silver print
(National Archives)

In the fourth photograph we see the next stage of the relief process. Sacks of flour are piled high next to the Moulder Warehouse in the China Basin area, which was used by the Army as one of the major distribution points for south of Market neighborhoods. A fully ladened wagon awaits unloading by civilian day laborers. Military security is present and an American flag designates that this is an official government site. Shooting at a forty-five–degree angle, Givens not only captures the geometry of the piles of food sacks but also of the scatterings of rectangular timber in the left foreground.

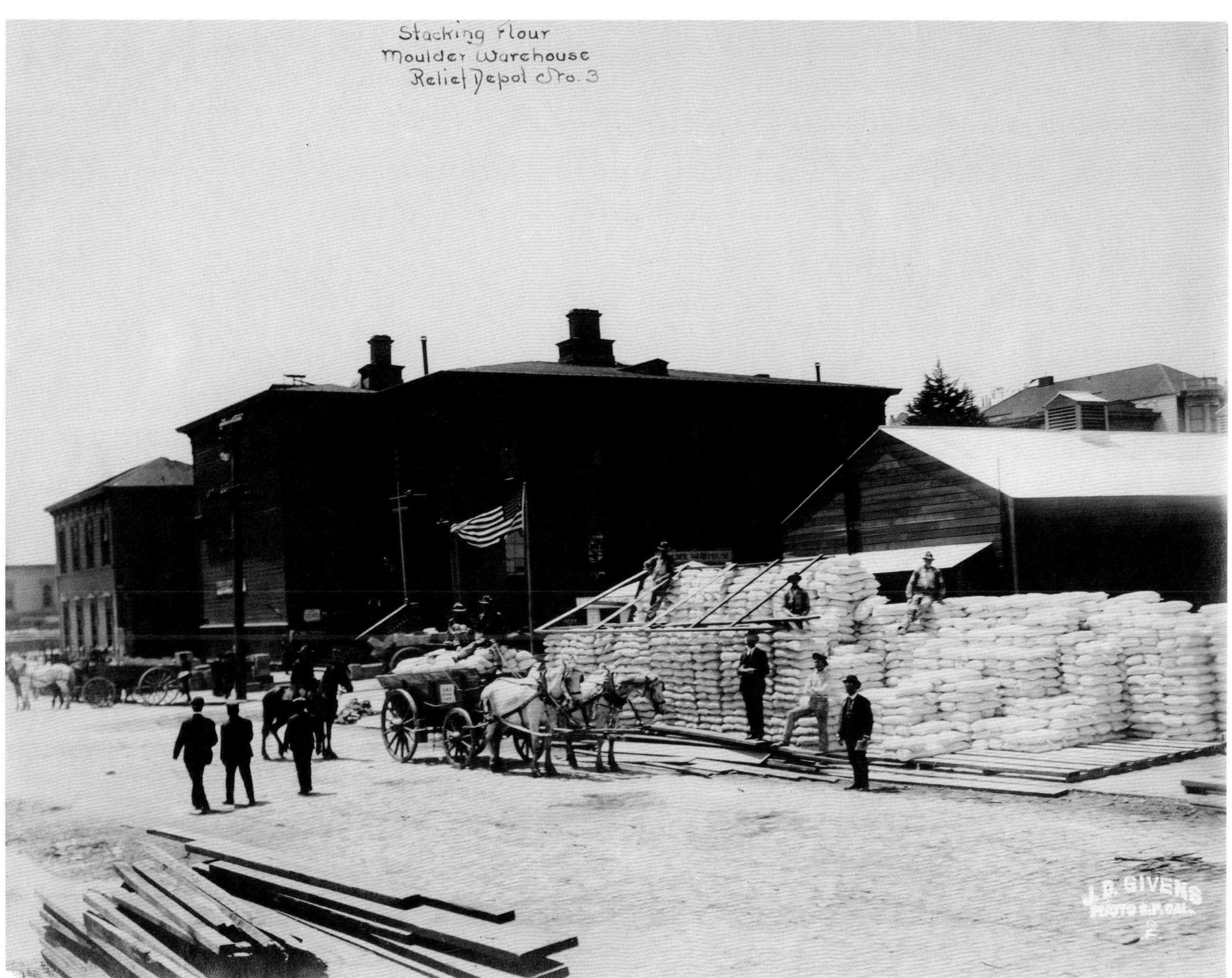

Stacking Flour
Moulder Warehouse
Relief Depot No. 3
J.B. GIVENS
PHOTO S.F. CAL.
2

PLATE 39. James D. Givens, *Moulder Warehouse Relief Depot No. 3,* 1906, gelatin silver print (National Archives)

The fifth photograph in the series hearkens back to Givens's earlier panoramic views of downtown San Francisco in the immediate wake of disaster. Here, we look north toward the top of Rincon Hill from China Basin in the city's southern sector. The camera scans a three block view of a region of San Francisco that was not damaged by fire. Standing, we can assume, on the stacked piles of potato bags in the foreground, Givens captures on the left an earthquake damaged structure, the classical revival Moulder Warehouse at Second and Townsend, a series of Victorian houses, and larger buildings at the peak of the hill. On the right appear a field of rubble, stacks of flour bags, a group of Eastlake houses and the rooflines of large homes atop Rincon Hill at a location now occupied by the approach to the Bay Bridge. The scene is completed by the depiction of quotidian activity taking place on the street in front of the food distribution depot.

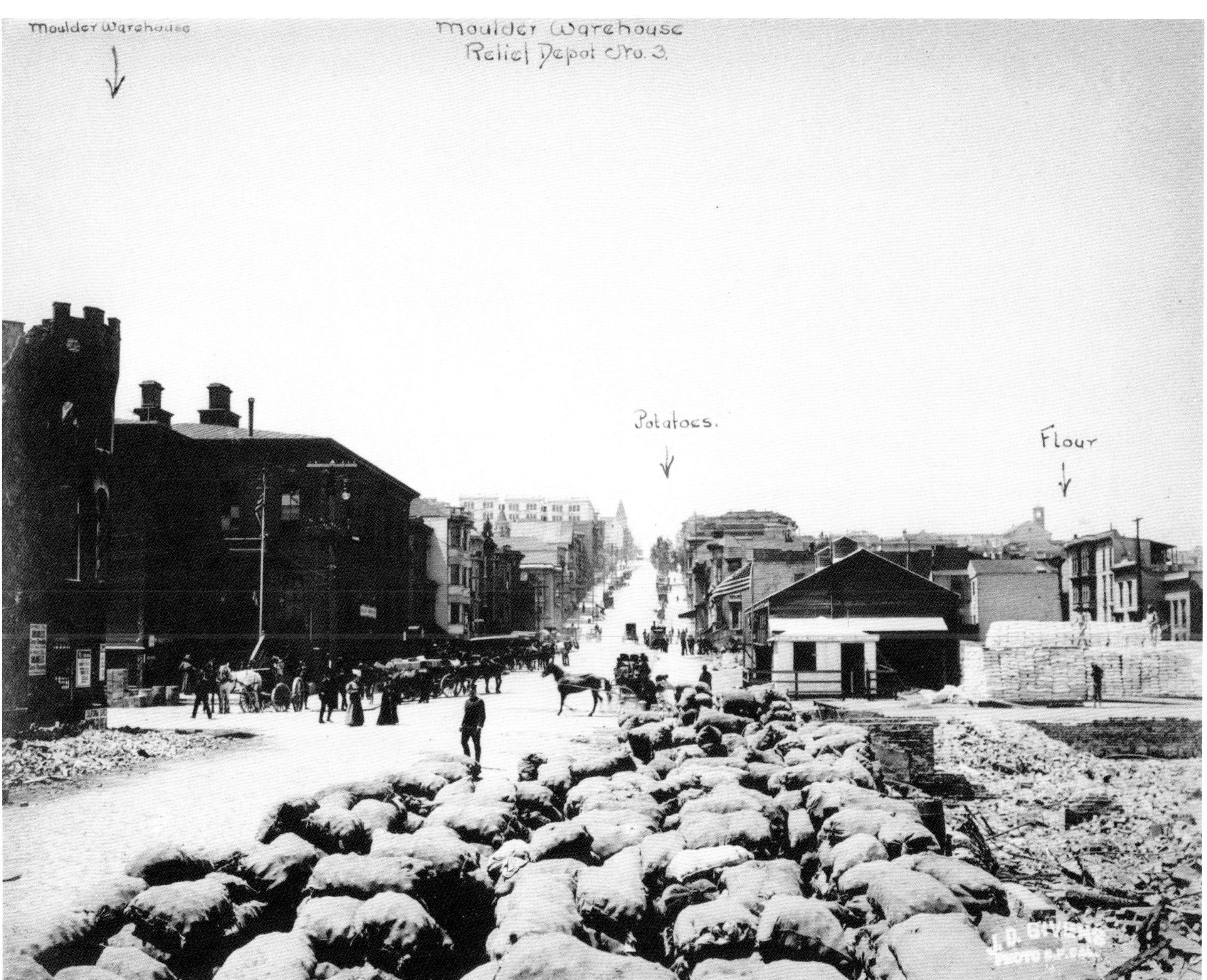

Moulder Warehouse
Moulder Warehouse
Relief Depot No. 3.
Potatoes.
Flour

Part-time photographer D. H. Wulzen stands at Battery Street looking westward along Bush Street as it rises along the heights of Nob Hill in the distance. The single, rather Chaplinesque figure with derby hat and what appears to be a doctor's bag striding amid the broken buildings and piles of rubble hearkens back to themes expressed in many photographs in the weeks after April 18: devastation and isolation in the streets of downtown San Francisco. Though great amounts of masonry rubble have been shoveled clear in the nearer blocks, cleanup still remains to be undertaken from Montgomery Street west. The edifice looming over this view is one of the true iconic structures in the pre-earthquake city, the Mills Building at Montgomery and Bush. Designed in restrained Romanesque style for banking millionaire Darius Ogden Mills in 1891 by nationally acclaimed Chicago architects Burnham and Root, it was the first building in California to be framed wholly in steel and was clad in brick walls with terra cotta decoration. Though badly damaged by fire, this Chicago-style skyscraper, a touchstone for many other similarly styled buildings in the city, was reopened less than two years later having been enlarged and with more generously decorated facades.

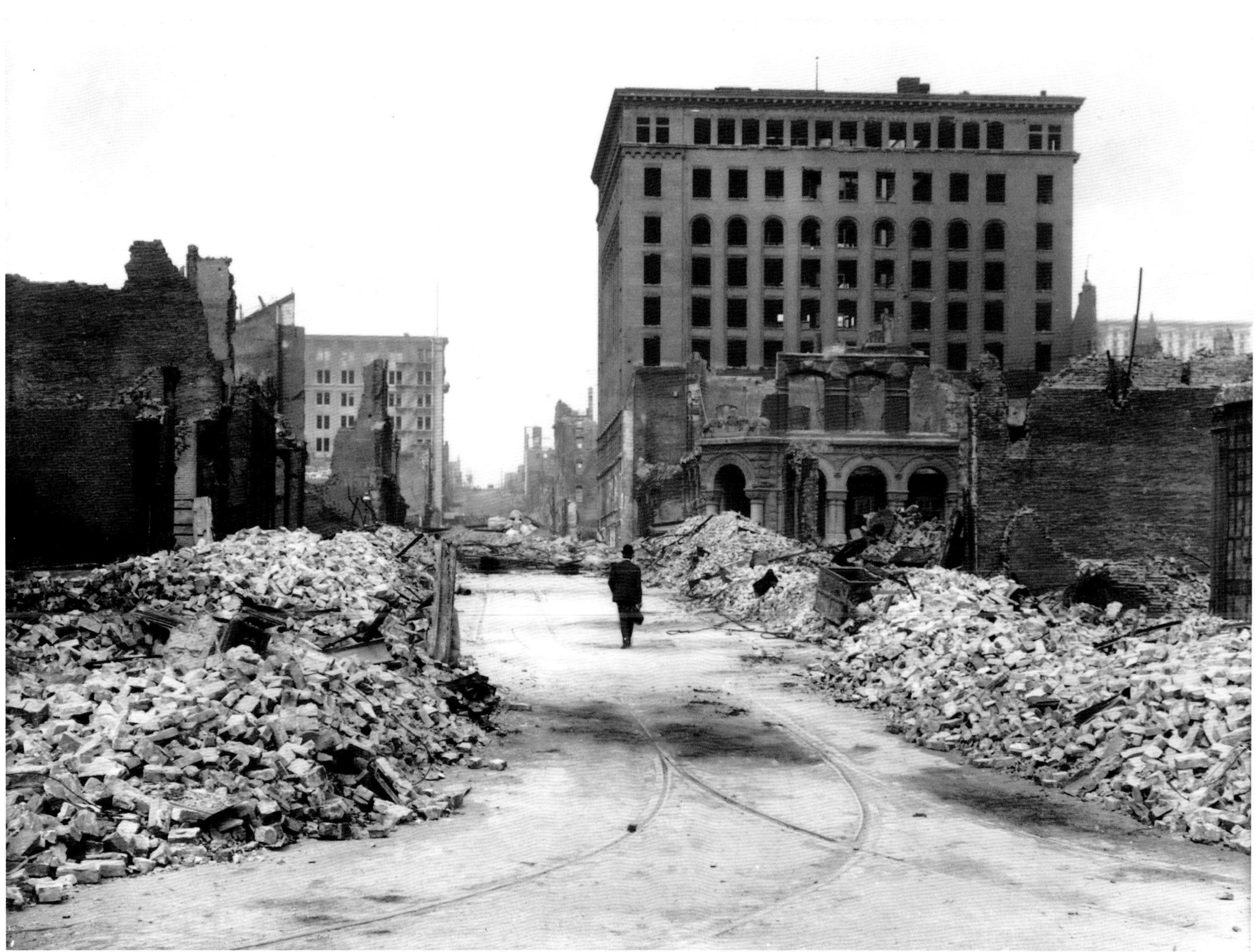

Life slowly returns to downtown San Francisco in this Wulzen view taken from Sansome Street looking west along Sutter. The combination of sharp focus objects in the foreground and soft focus ones in the distance lends an almost Hollywood set-like quality to the scene. Signs of destruction are everywhere: rubble in the foreground, the bent, de-nuded building frame on the right, broken brick buildings in the middle ground, and, in the distance on the right, one of two great towers flanking the entrance to the massive Temple Emanu-El at 450 Sutter. Consecrated in 1866, this house of worship was the religious home for the city's liberal Jewish community. Its twin towers were crowned by gigantic onion shaped domes that were destroyed in the earthquake. They were the highest, as well as the most exotic sights in the Union Square area. The wholesale destruction depicted in views such as this makes it all the more amazing that downtown San Francisco was largely rebuilt—from Market to Nob Hill, from the Embarcadero to Van Ness—in only three years.

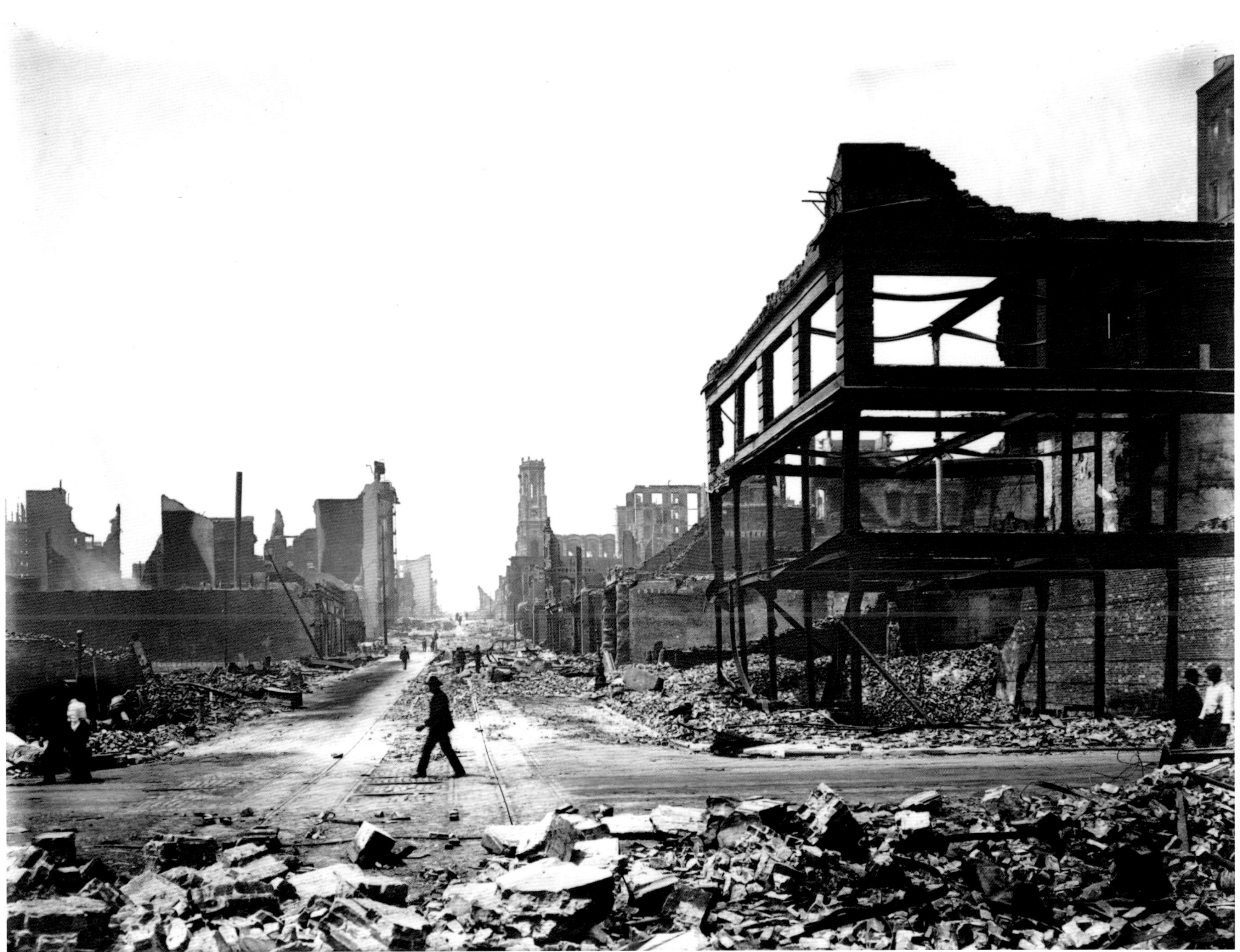

A cavalryman from the Presidio casually rides across Dupont Street (Grant) days after the disaster has wrought devastation on the eight square blocks of Chinatown. Though Nob Hill was made of firmer soil than many areas of the city, the flimsy construction of buildings in this neighborhood spelled doom for tragic numbers of Chinese inhabitants. Pointing his camera south along Dupont from near Clay Street, Wulzen captures the gaping ruins of San Francisco's beloved old St. Mary's church, which was consecrated in 1854. The Gothic Revival brick structure with its bell tower, its pointed arch windows and portals, its buttresses, and its stained glass windows was the first European-scale basilica church built in the boomtown. It was unthinkable to the large local Catholic community that it not be rebuilt.

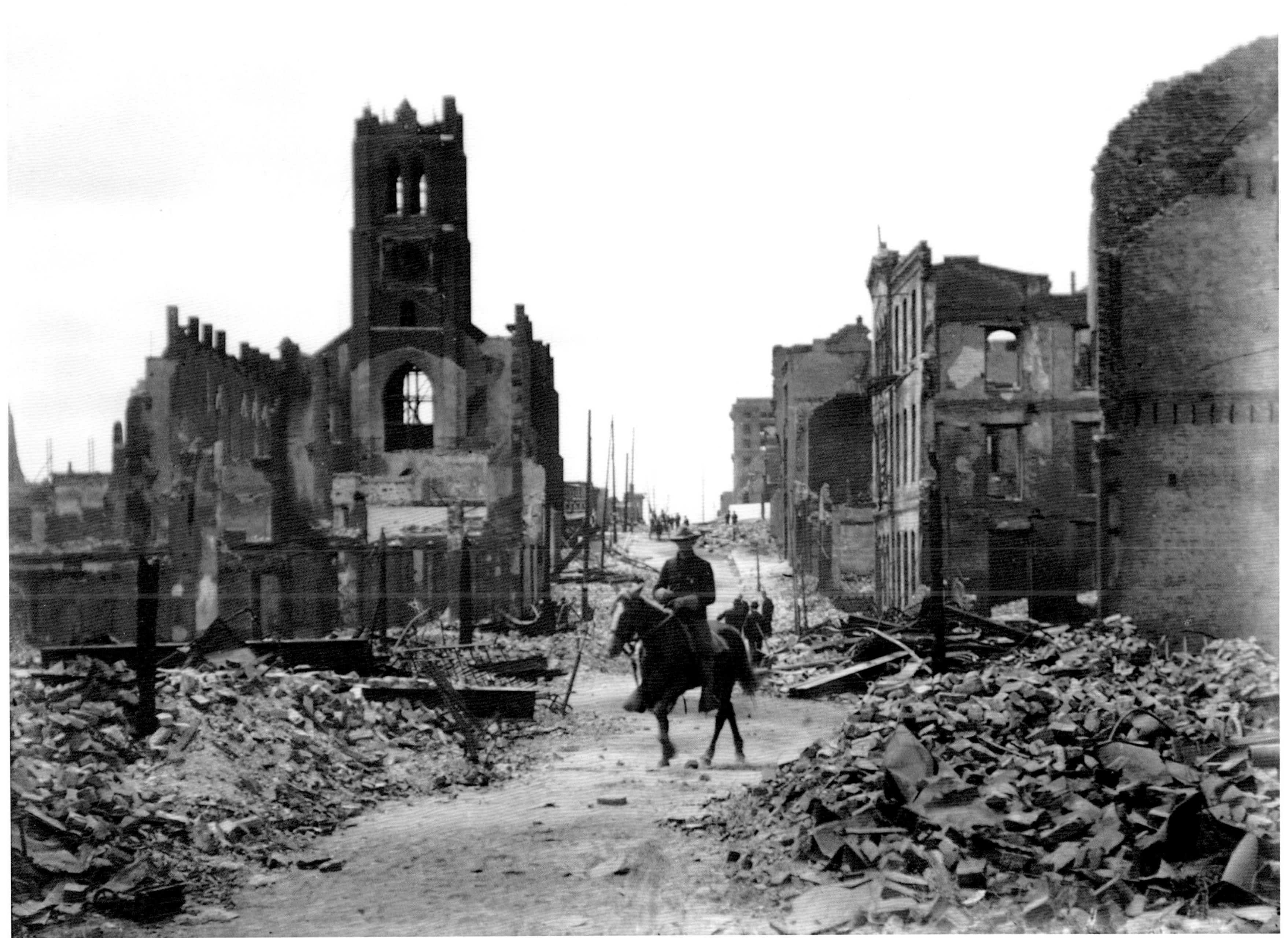

PLATE 43. D. H. Wulzen, *Untitled,* n.d., glass plate negative
(D.H. Wulzen Collection, San Francisco History Center, San Francisco Public Library)

Wulzen positioned his camera on upper Market Street near Seventeenth in order to capture this panoramic view of the badly damaged downtown area. In the lower right corner we see at close range the damage done by the earthquake to a large brick building and adjoining tower. In the middle distance to the left are the skeletal remains of City Hall. This most obvious symbol of San Francisco's destruction would remain visible until its demolition three years after the disaster. Across Market from City Hall stands the new Federal Post Office building which was near completion in April of 1906. Further down Market Street are visible the Emporium department store, the Call Building and the Ferry Building, to the right of which rise Rincon and Potrero Hills. The shadowy presence of the East Bay hills looms on the horizon. The four city streetcars visible in the photograph imply the return to normality of part of the urban infrastructure.

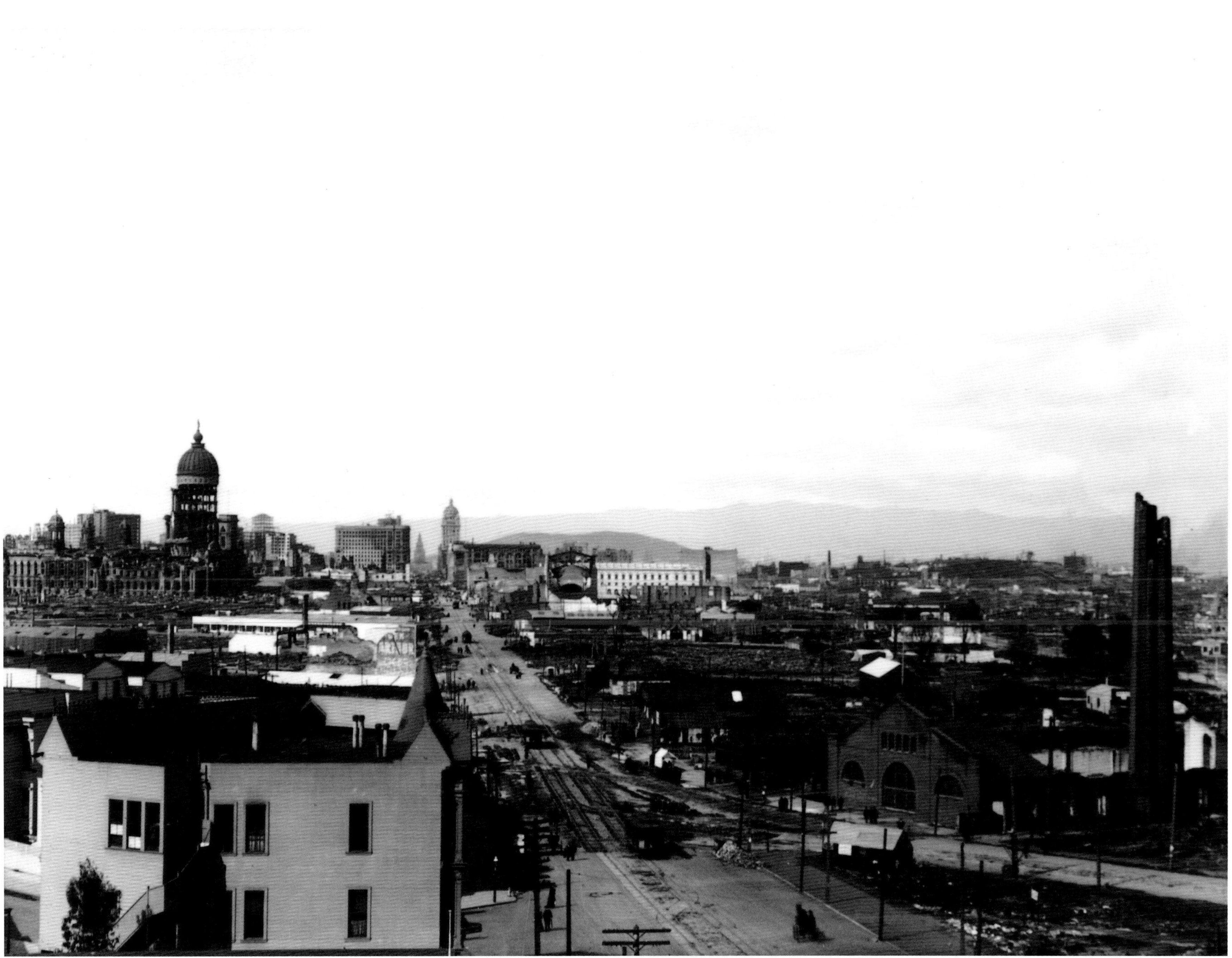

In this view Wulzen turns his camera away from the destruction of the downtown area and aims it west from upper Market Street toward Twin Peaks. He documents one of the instant tent villages that arose on sites west of Van Ness Avenue in the days after the fire was extinguished. This small temporary settlement was better located than many because of its proximity to an existing and relatively undamaged neighborhood where goods and services could be procured. Despite this fact, Wulzen's distancing view expresses a kind of forlorn atmosphere with images of irregularly positioned tents, barren rock outcroppings, the emptiness of activity on Market Street, and the insignificance of human figures within a grand panoramic vista.

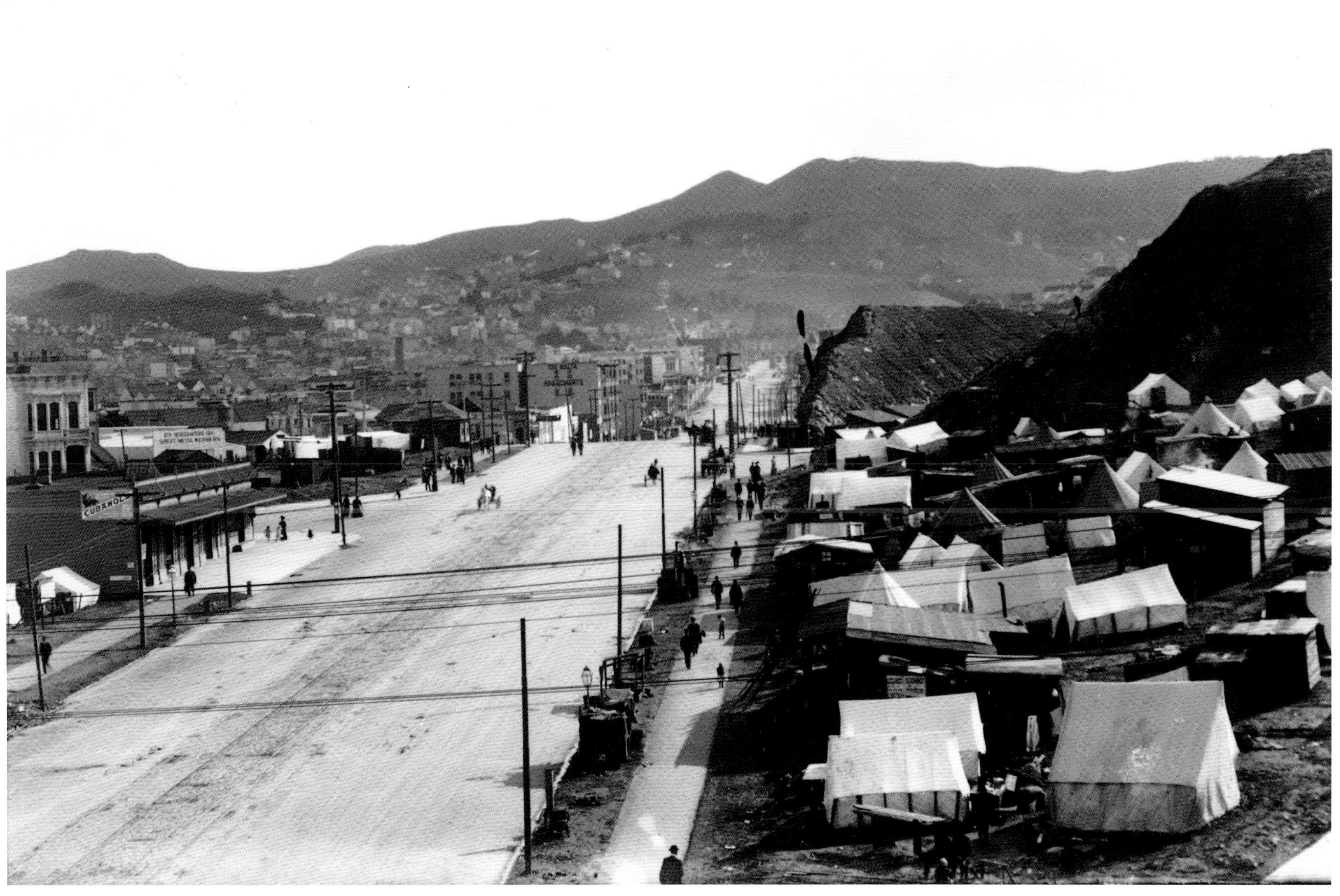

CUBANOL

In this photograph Wulzen contrasts destruction with resurrection. He stands near the corner of Third and Howard Streets looking north toward Market. In the foreground poses a nattily attired man whose posture echoes the diagonal axis of the broken metal girder to his right. Across Howard are the facades of buildings mortally wounded by the forces of nature. In the distance stands the largely undamaged exterior of the Call Building, flanked by new edifices under construction. The unfinished steel-framed structure to the left of the Call is the Whittell Building designed by Shea and Shea and begun the year before the earthquake. Still standing at 166 Geary, the Whittell suffered little damage from the tremor due to its heavily reinforced steel frame and was opened for occupancy later in 1906. The combination of decimation in the foreground and renewal in the background is a powerful and succinct statement about the city's response to disaster and thus about its prospects for the future.

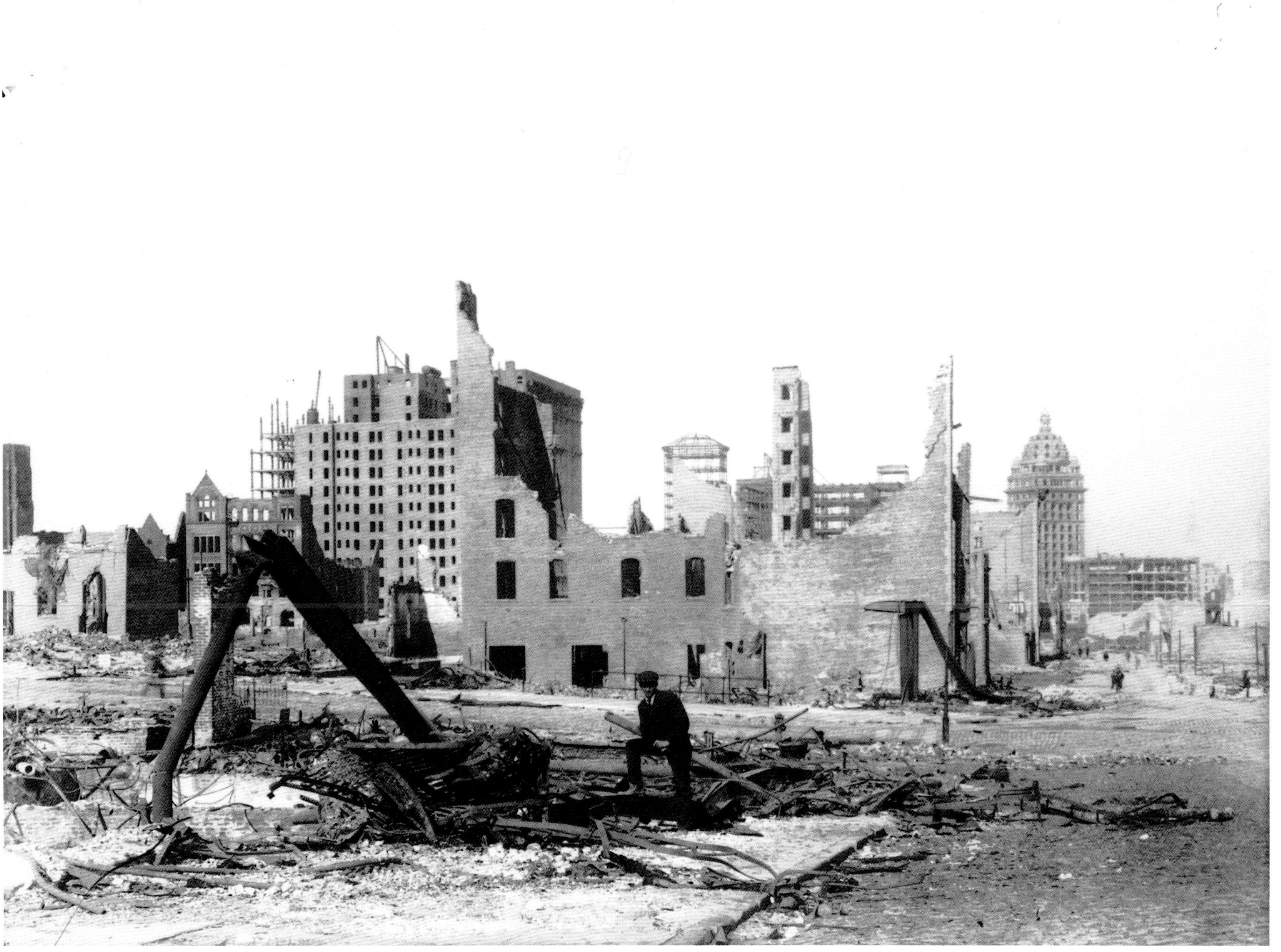

Photographer D. H. Wulzen owned the pharmacy at Seventeenth and Castro Streets in the upper Market Street area depicted here. In the weeks after April 18, his place of business served as a temporary food distribution center and post office for the residents living in the surrounding neighborhoods. Though there does appear to be some minor damage to the masonry church building to the right, the impressive wooden structure that housed Wulzen's establishment—with its Eastlake decorations, bay windows, and decorative iron balustrade still intact—provides evidence that the destructive power of the earthquake did not exert itself in all parts of the city. The technical quality of this view is excellent though its significance probably lies in a combination of reportage and self-interest on the photographer's part, not in larger pictorial and symbolic themes.

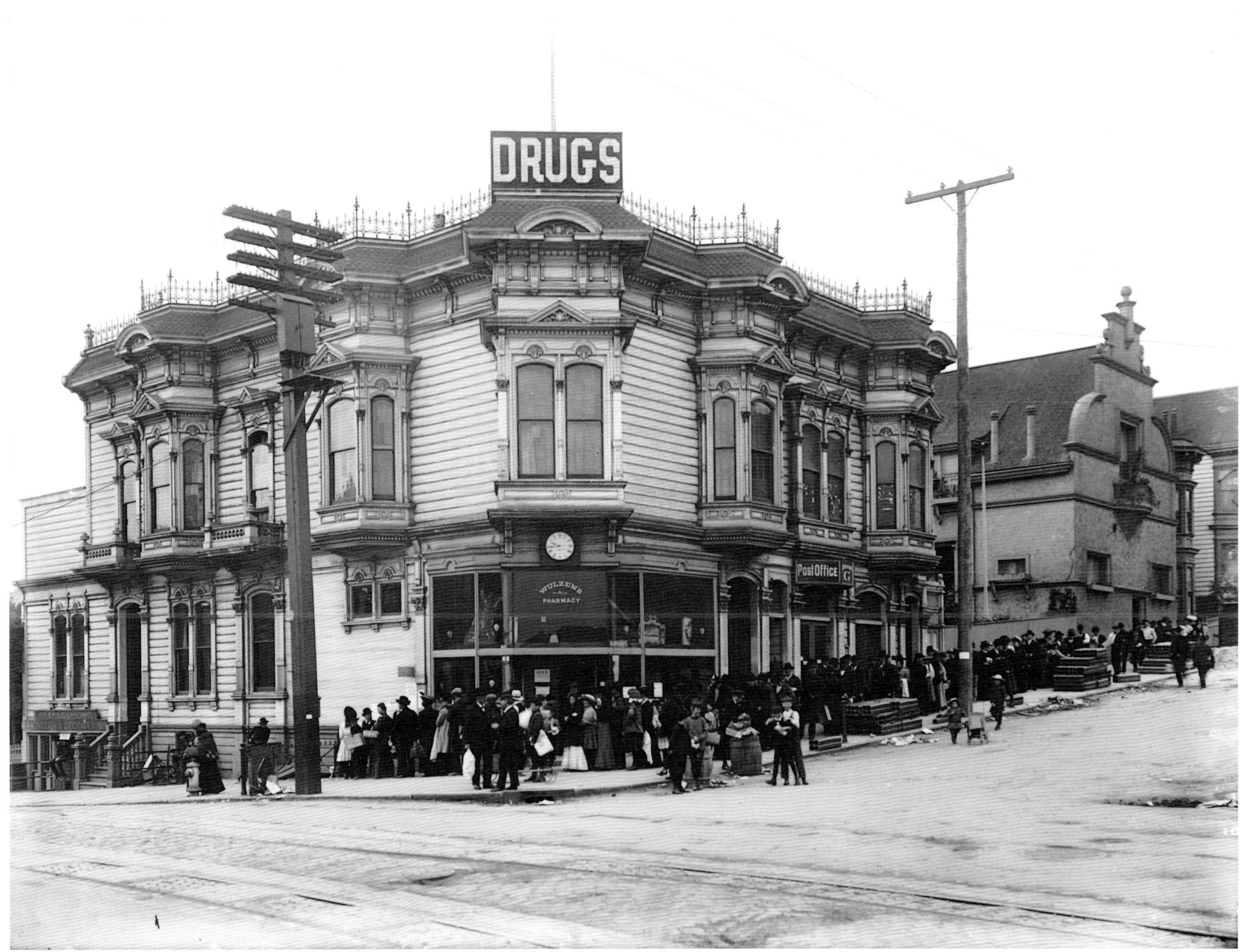

DRUGS
WULZEN'S
PHARMACY
Post Office

PLATE 47. D. H. Wulzen, *Untitled,* n.d., glass plate negative
(D.H. Wulzen Collection, San Francisco History Center, San Francisco Public Library)

Within a day after the earthquake, city edicts were issued forbidding San Franciscans to cook on their kitchen stoves. Many fires had been started by embers falling on wooden roofs whose chimneys had collapsed during the tremor. In this view we see Wulzen's wife preparing dinner in her makeshift kitchen constructed out of three doors, a wood stove, and a jerry-rigged exhaust carrying fumes and embers out through a nearby window. Among the Victorian staples visible here are the cast iron stove, the heavy metal kettle, the pressed back chair and the detailed oak doors. Again, the purpose of this photograph seems to be twofold: documenting the personal experiences of the photographer and recording the temporary conditions of life in the weeks after the disaster. The composition of the view is quite competent, framing Mrs. Wulzen between two rectangular doors on a diagonal bias.

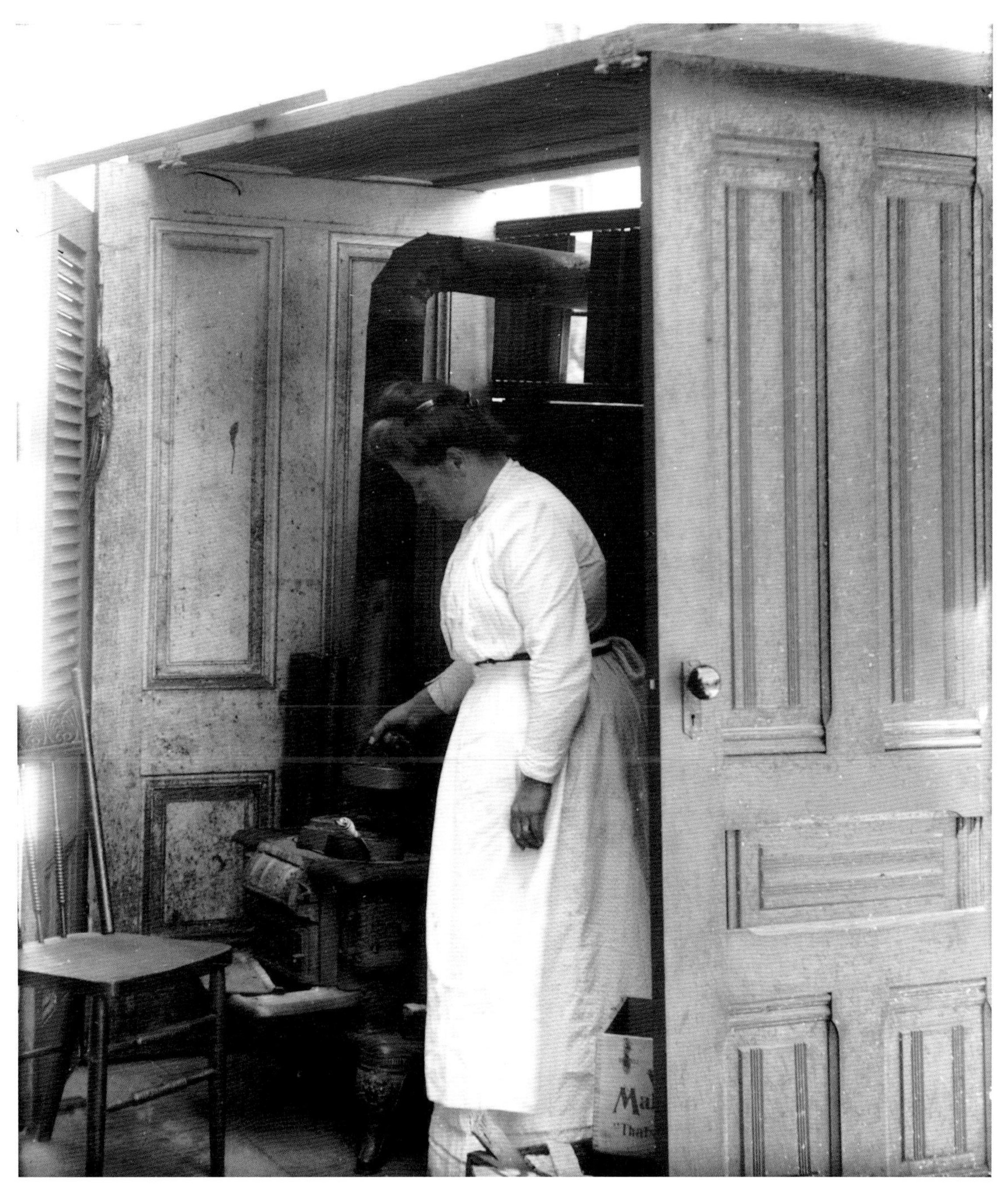

Willard Worden provides a rare glimpse into the less than admirable behavior of some San Franciscans during the city's hour of need. Though most photographers stressed the orderliness, heroism, and perseverance of the beleaguered population, there were episodes of looting and lawlessness of which around a dozen cases resulted in death by firing squad under the terms of martial law. Here, two looters are made to stand before the camera holding up signs reading "I am a junk thief" and "So am I," while a unit of federal troops with rifles at hand witnesses the embarrassing admissions of guilt. Worden's treatment is none too subtle in its expression of severe consequences for antisocial behavior and moral condemnation of such acts. The photograph is taken on the cobblestone road running along the city's Embarcadero waterfront, just north of the Ferry Building. The latter, which had opened in 1898, escaped the great fire but was superficially damaged during the earthquake. The scaffolding around the building indicates that this view was taken some weeks after the city's destruction.

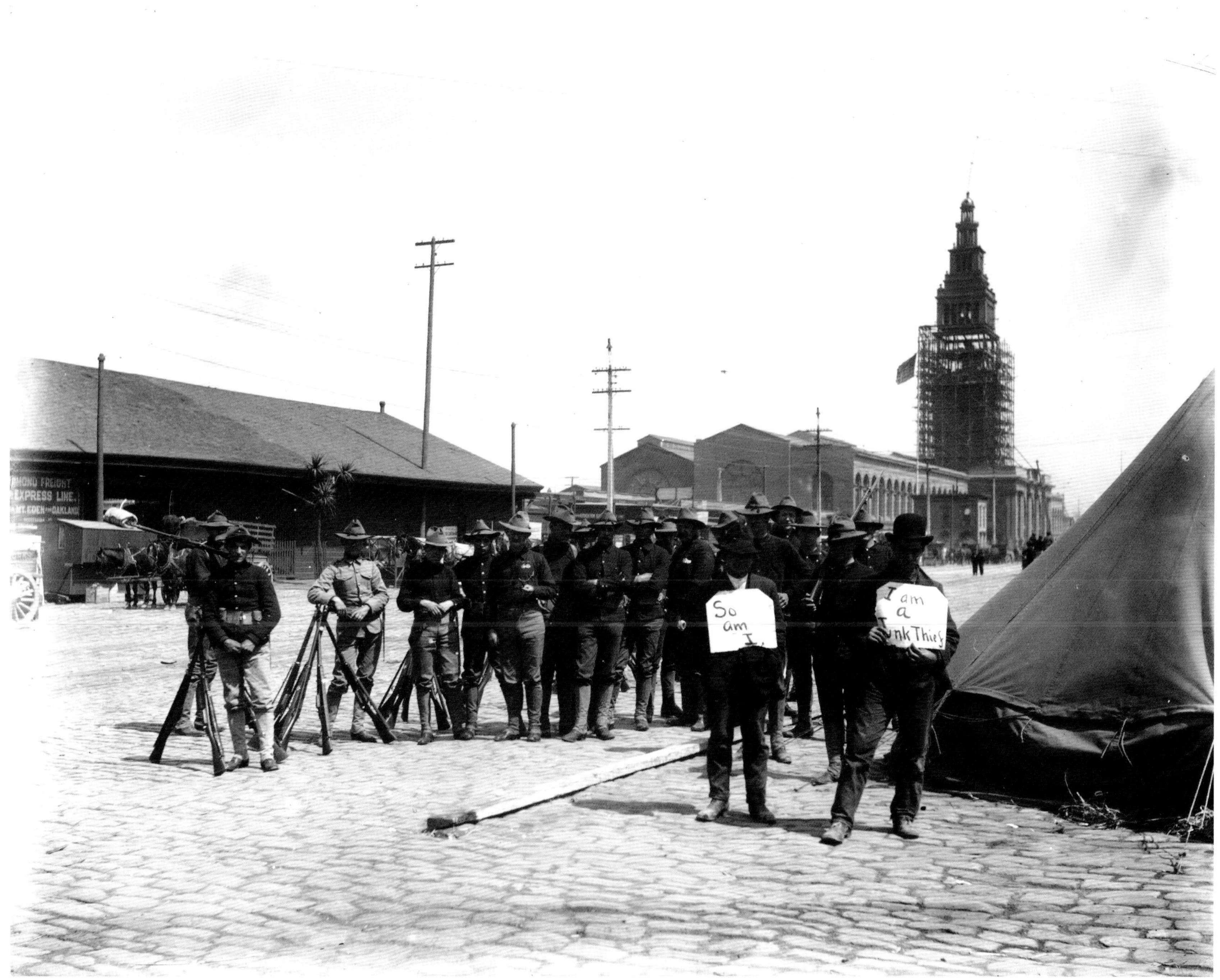
So am I.
I am a unk Thief

Stellman published visual essays on San Francisco's "vanished ruins," postearthquake Chinatown, and the Panama Pacific International Exposition of 1915. Though a journalist by profession, Stellman's pictorialist aesthetic sometimes uncomfortably combines impressionistic reportorial detail with a nostalgia that borders on the saccharine. In this view down the slope of Nob Hill along California Street above Stockton we see downtown San Francisco acting as a backdrop for the evocative images of two destroyed brick Gothic churches on the right and the heavily damaged old St. Mary's at the middle left. The large, gutted bell tower that occupies the center of the photograph belongs to Grace Cathedral, which had been dedicated in 1862 by Episcopalian Bishop Ingraham Kip. After its destruction the new Grace Cathedral was built where it stands today—on California Street between Taylor and Jones—on land donated by William H. Crocker. The most suggestive features of this photograph are the fragmented houses of religion that dominate the foreground.

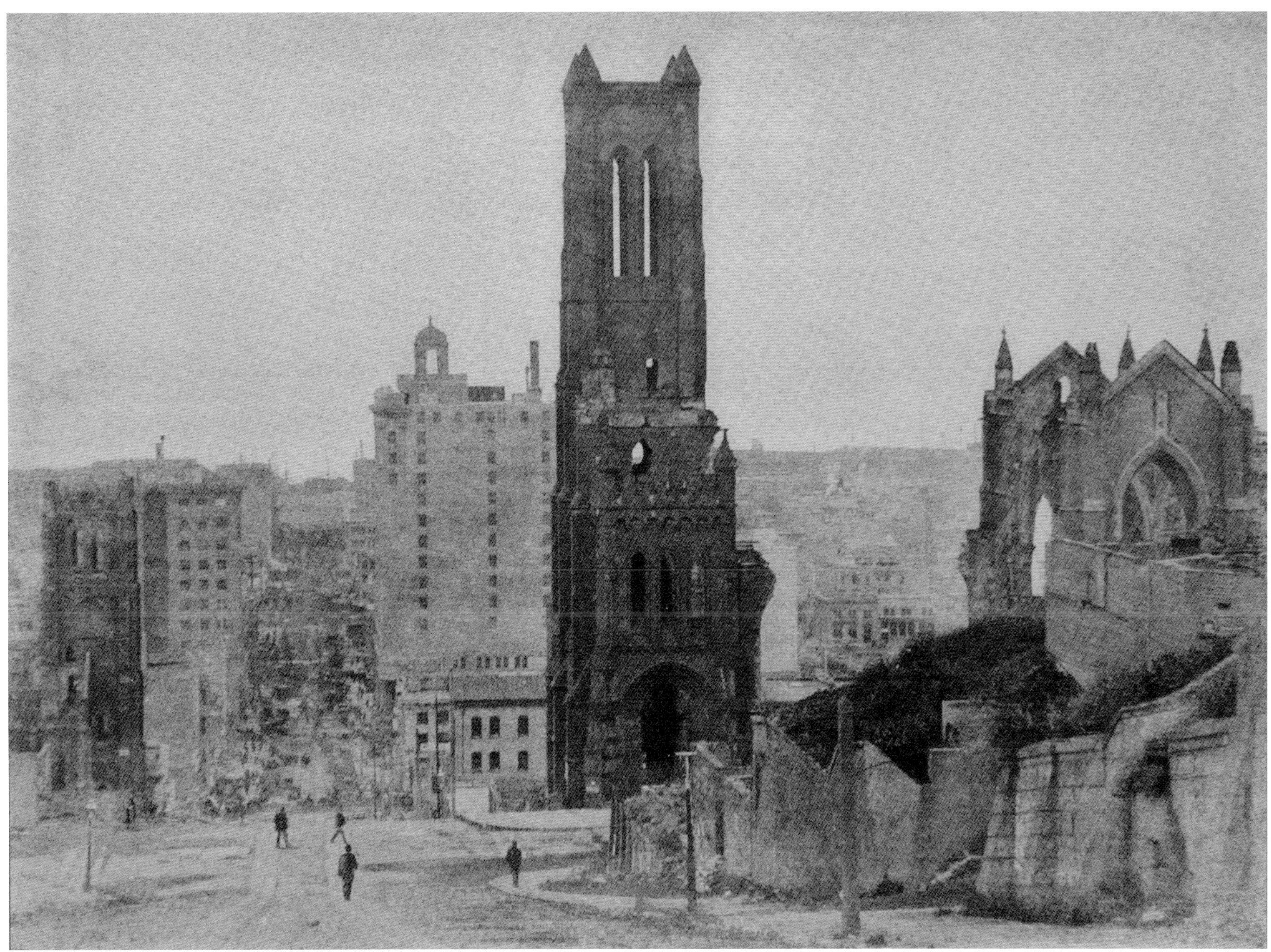

PLATE 50. Louis J. Stellman, *Untitled,* 1910, photogravure (private collection)

This photograph, published in Stellman's book *The Vanished Ruin Era* (1910), depicts one of many government and Red Cross run camps that sprang up on vacant lots and in city parks (including Golden Gate Park) west of Van Ness. From April to June some 8,000 identical prefabricated "refugee houses" were built with government money and private contributions. Some dispossessed San Franciscans lived in these conditions for months and even years. In many post-disaster photographs the long, unbroken rows of uniform structures implied both the continuity of civilized social order and the loss of self-determination. Stellman seems to be depicting neither of these symbolic implications here. His camera captures only a small number of houses and includes no human images to give any distinctiveness to the scene. Such views as this must be regarded as mere photographic cataloguing.

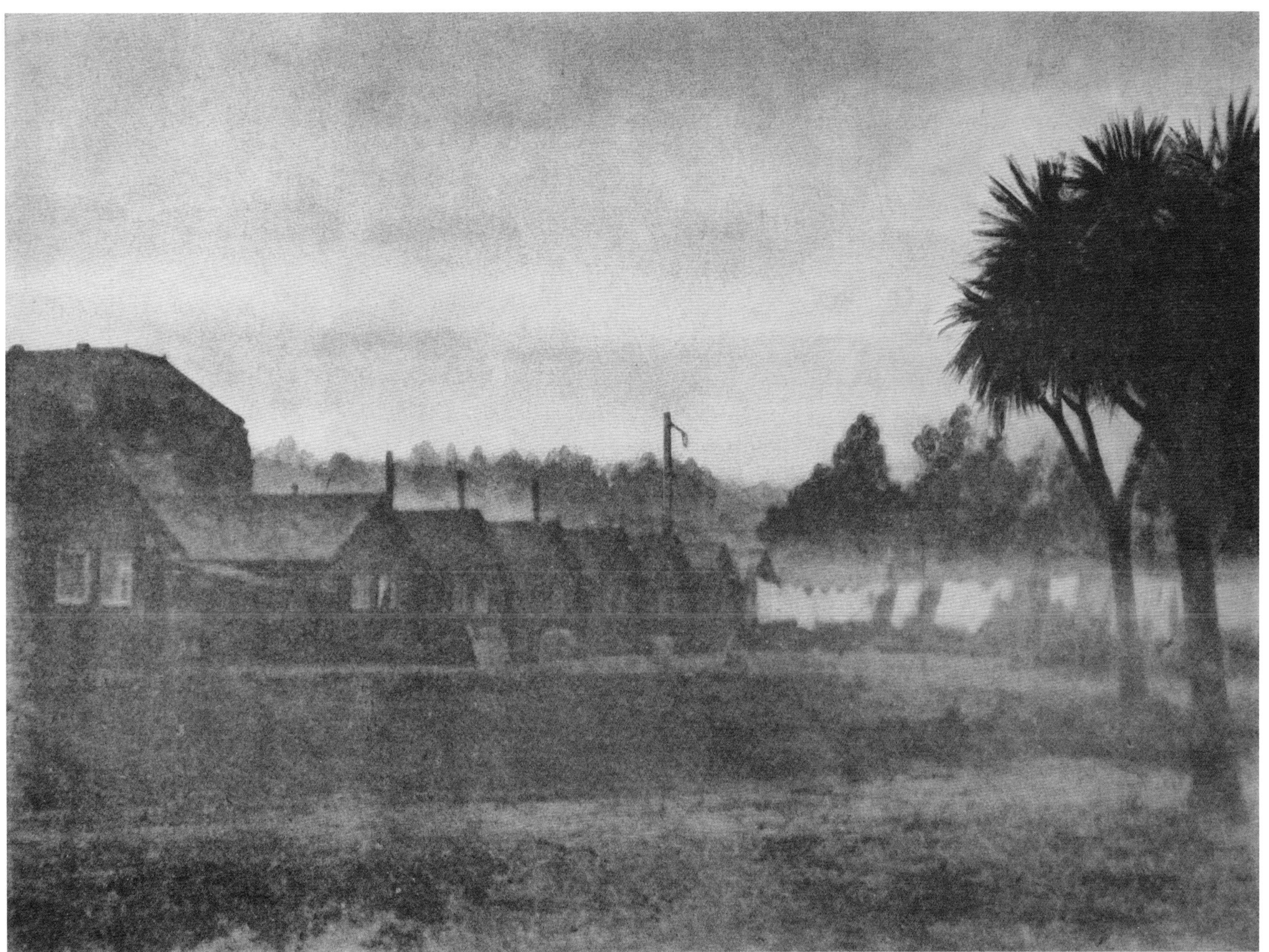

This photograph, taken by Stellman in 1909, depicts the broken twenty-one-foot tall statue of the Goddess of Progress that fell to the ground when the cable lowering it broke during the demolition of City Hall. For three years after the earthquake and fire, the Goddess, holding the torch of freedom in her raised right hand, had stood undaunted atop the largely destroyed City Hall tower. Now, her accidental demise offered Stellman the opportunity to record a story of local interest and make a statement about human vanity at the same time. The broken torso with raised right arm lies upon a pile of rubble with its head opposed to its body. A curiosity to be sure, but one which evoked memories of the disaster three years earlier.

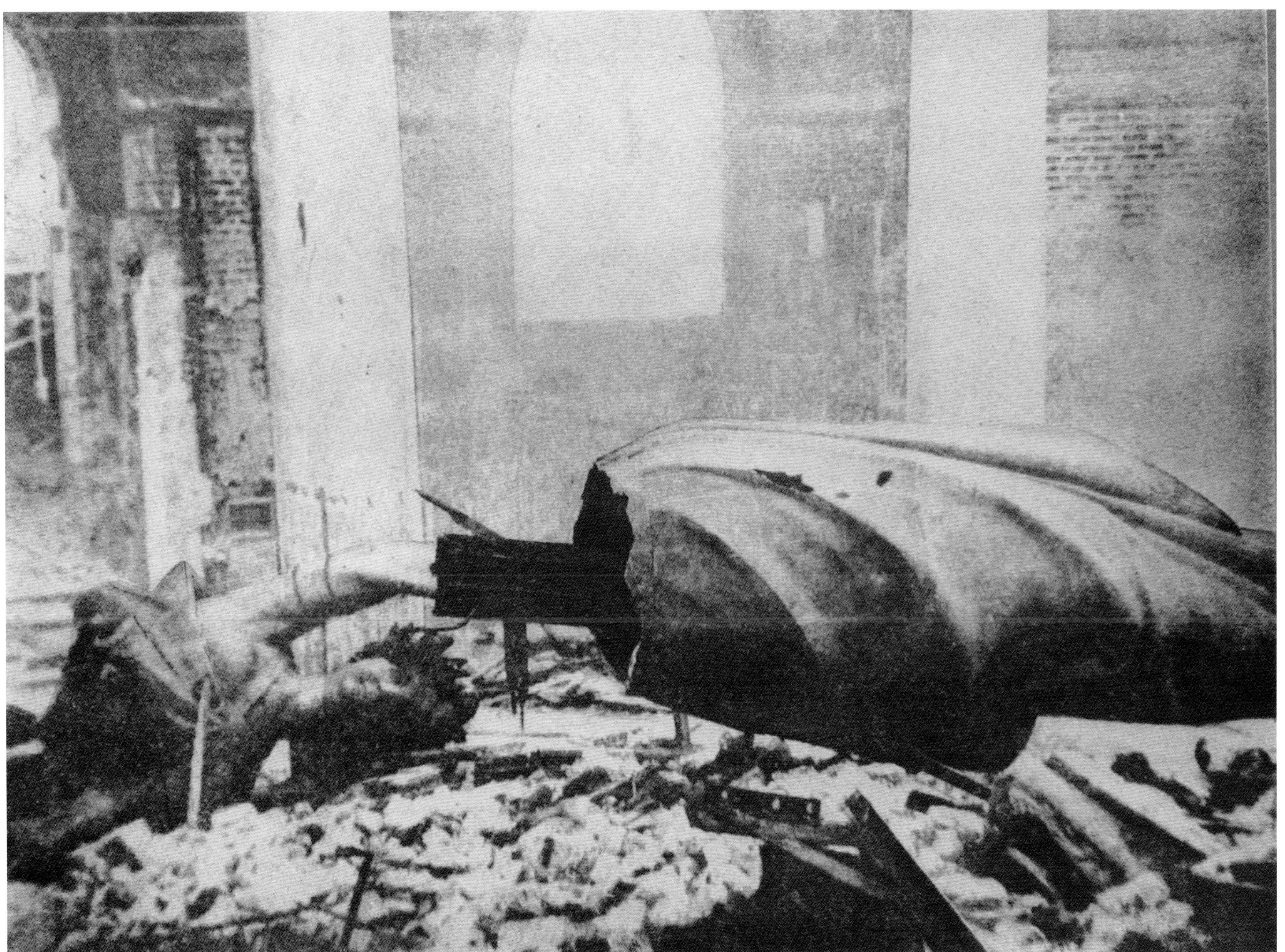

One of the first major retail and office buildings to be started on Market Street after the 1906 disaster was the Phelan Building, which still stands at the confluence of Market and O'Farrell Streets at Grant. Erected on the site of an earlier Phelan Building, it was commissioned by the brilliant young banker James Duval Phelan who had served as the city's progressive Democratic mayor from 1897 to 1901 and would later be elected to the United States Senate from the state of California. Phelan hired respected architect William Curlett to design a ten-story flatiron-style structure climaxing in a rounded corner engineered with such support strength that an additional ten stories could be added on at a later date. Construction at the site progressed from 1907 to 1908 and the R. J. Waters Company photographed the new Phelan as it rose from street to completion in September of the latter year. In this photograph looking northwest from across Market Street, the powerful steel girders and supports of modern skyscraper construction outline the shape of the building. Such reinforced construction had been used with great success in San Francisco before the earthquake and was regarded as both temblor and fire proof. This view also offers some interesting contemporary political commentary. On the temporary wall surrounding the site is a series of signboards advertising not only various businesses and commodities but also touting the mayoral candidacies of two local political figures. In the middle of the wall on top is a sign supporting the Union Labor Party candidate P. H. "Pinhead" McCarthy. To the right are two signs supporting Daniel A. Ryan, a Republican who was a strong supporter of the corruption trials of "Boss" Abraham Ruef and deposed Mayor Eugene E. Schmitz, both of whom were currently in prison under indictment. This photograph was taken on November 5, 1907, the day of the general election in which both McCarthy and Ryan lost decisively to attorney, doctor, and poet Edward R. Taylor, the Democrat.

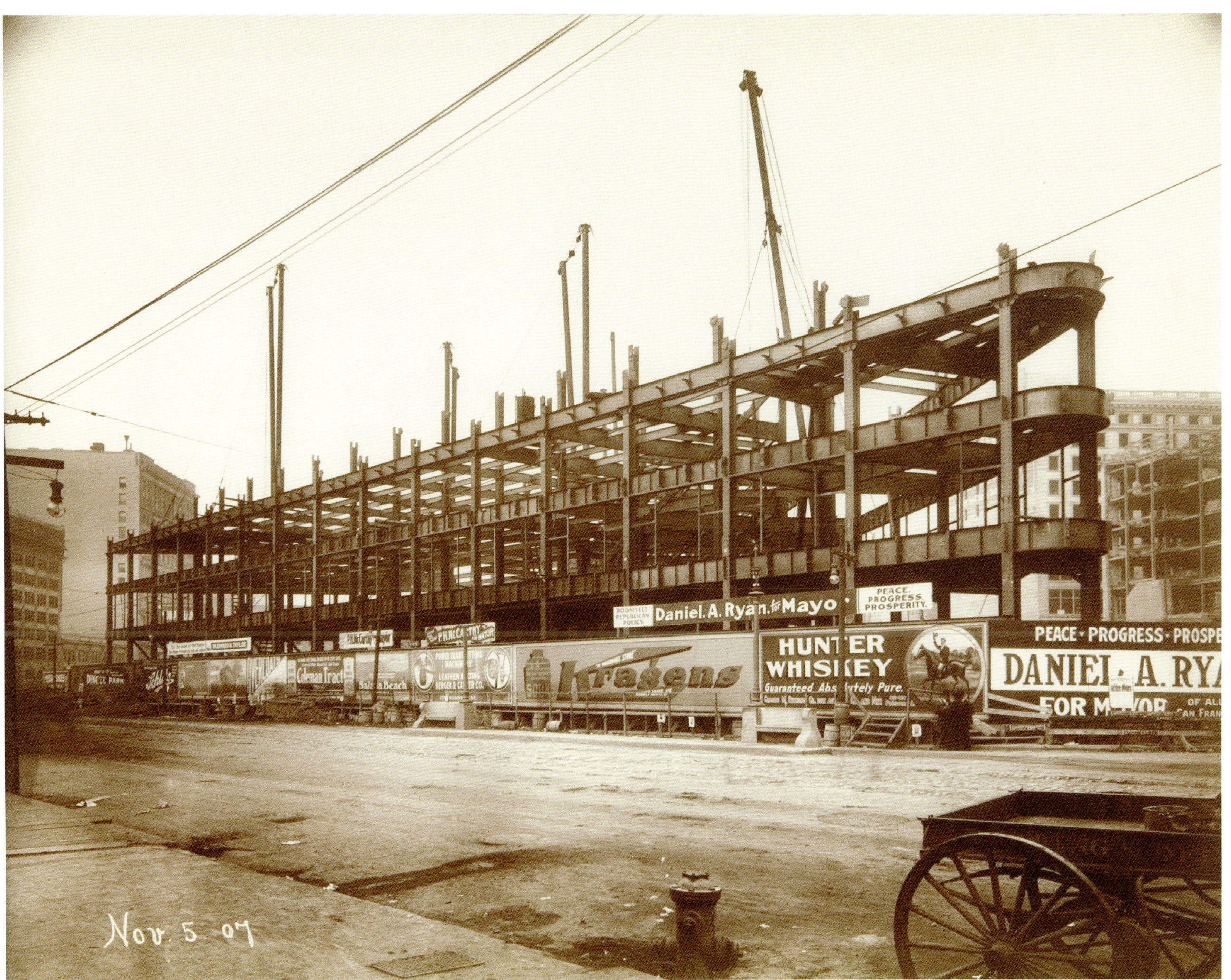
PEACE·PROGRESS·PROSPERITY.
Daniel. A. Ryan for Mayor
Roosevelt Republican Policy.
HUNTER WHISKEY
Guaranteed Absolutely Pure.
Kragens
Coleman Tract
Salada Beach
PEACE·PROGRESS·PROSPE
DANIEL·A·RYA
FOR MAYOR
Nov 5 07

The speed with which, by 1907, large buildings could be constructed using preformed steel members is evident in this Waters image. Two weeks after the previous view was made two more floors of structural framing have been added. Note the projections on the horizontal beams placed to support the building's sheathing and the large metal window frames permitting modern structures more access to light than was possible in thick walled earlier structures. The political campaign signs have come down, and the Hunter Whiskey advertisement now abuts a sign touting high-top True Merit Shoes.

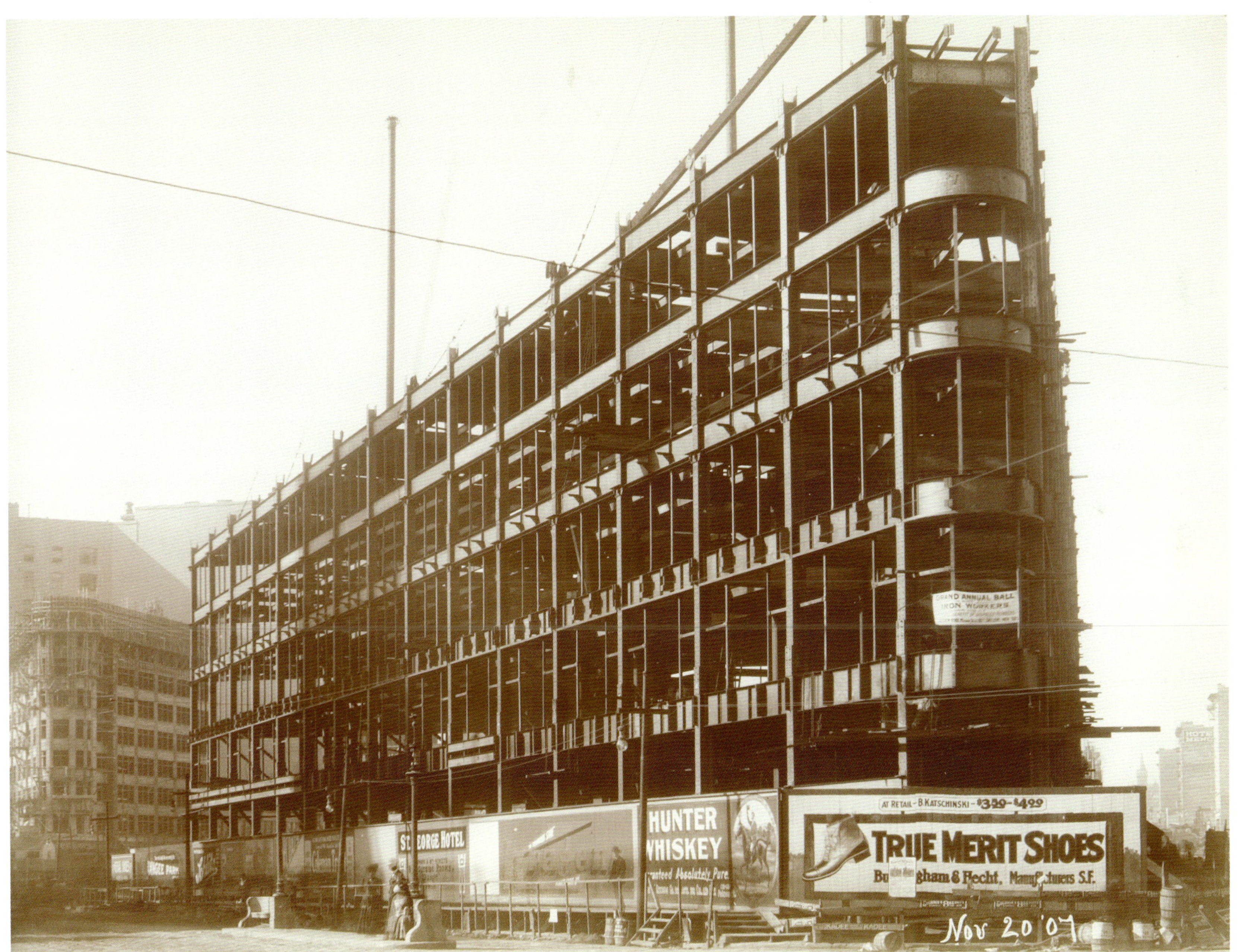

ST. GEORGE HOTEL
HUNTER WHISKEY
Guaranteed Absolutely Pure
AT RETAIL - B. KATSCHINSKI - $3.00 - $4.00
TRUE MERIT SHOES
Buckingham & Hecht, Manufacturers S.F.
GRAND ANNUAL BALL
IRON WORKERS
Nov 20 '07

This photograph, taken on April 3, 1908, shows the building's first seven stories now nearly completed and scaffolding covering the whole midsection and upper stories of the structure. Some of the decorative elements are now evident, including the broken rustication of the third story, inscribed window frames, and use of narrow columns to divide the spacious windows into tripartite sections. One major advantage of using terra cotta cladding rather than brick is that the former, a ceramic product, can be much more intricately modeled. Since heavy support walls were no longer needed in steel-reinforced construction, tall buildings could take greater liberties with the sheathing materials used on their exteriors.

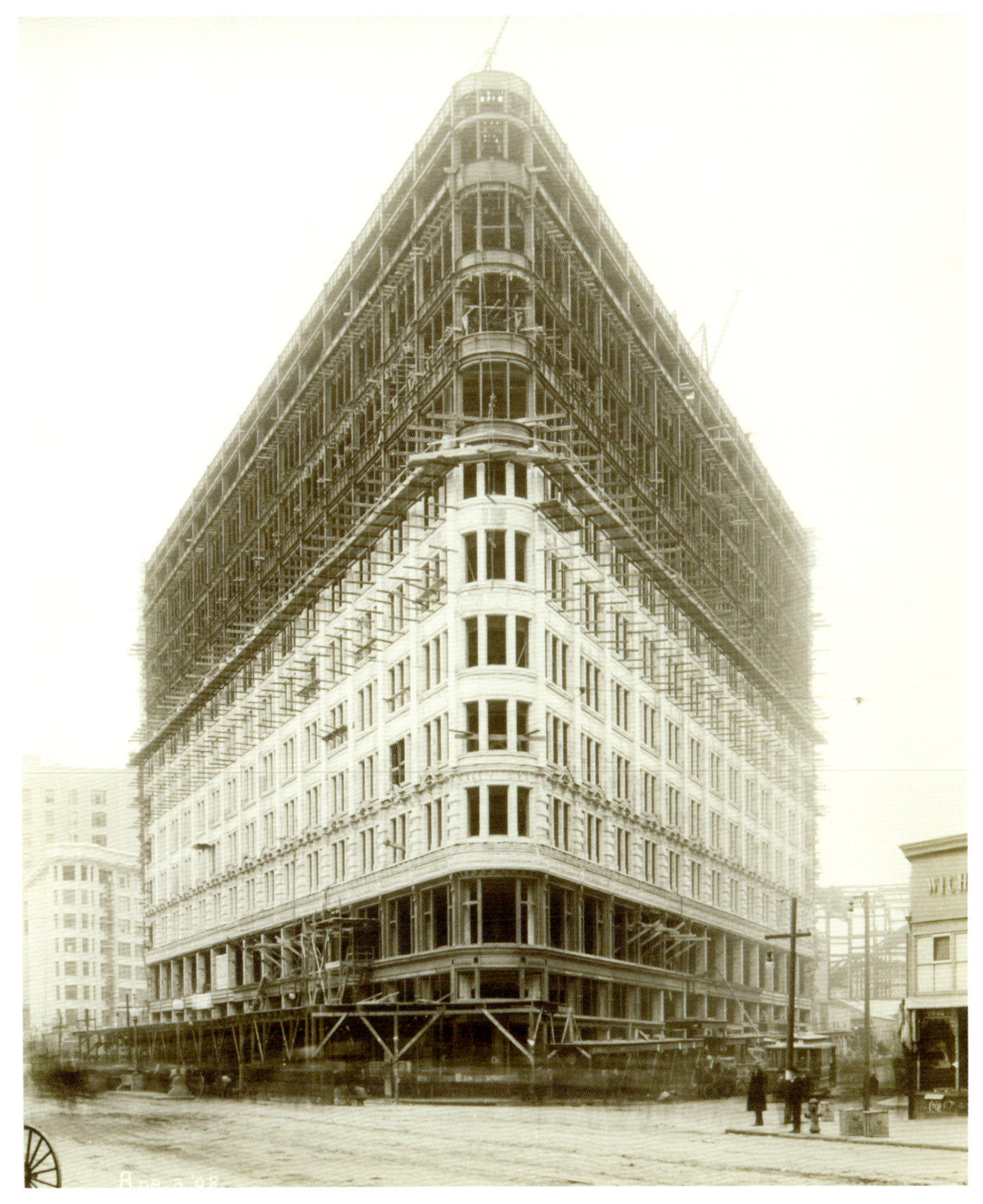

This final photograph taken by the Waters Company cameraman shows the Phelan Building as it looked on September 19, 1908. The retail spaces to which the first two glass walled floors are dedicated now show signs of life—with awnings attached to some of the entrances and a tobacconist doing business on the second story at the corner. As a final gesture of urban sophistication architect Curlett has attached a semicircular, Tiffany-esque glass awning over the main entrance at the building's corner under which visitors passed on their way to elevators taking them to the offices on the eight upper floors. What the photographer's lens has recorded here, in September 1908, may be seen as an impressive symbol not only of San Francisco's swift recovery from cataclysmic disaster but also of its place as the business capital of the western United States.

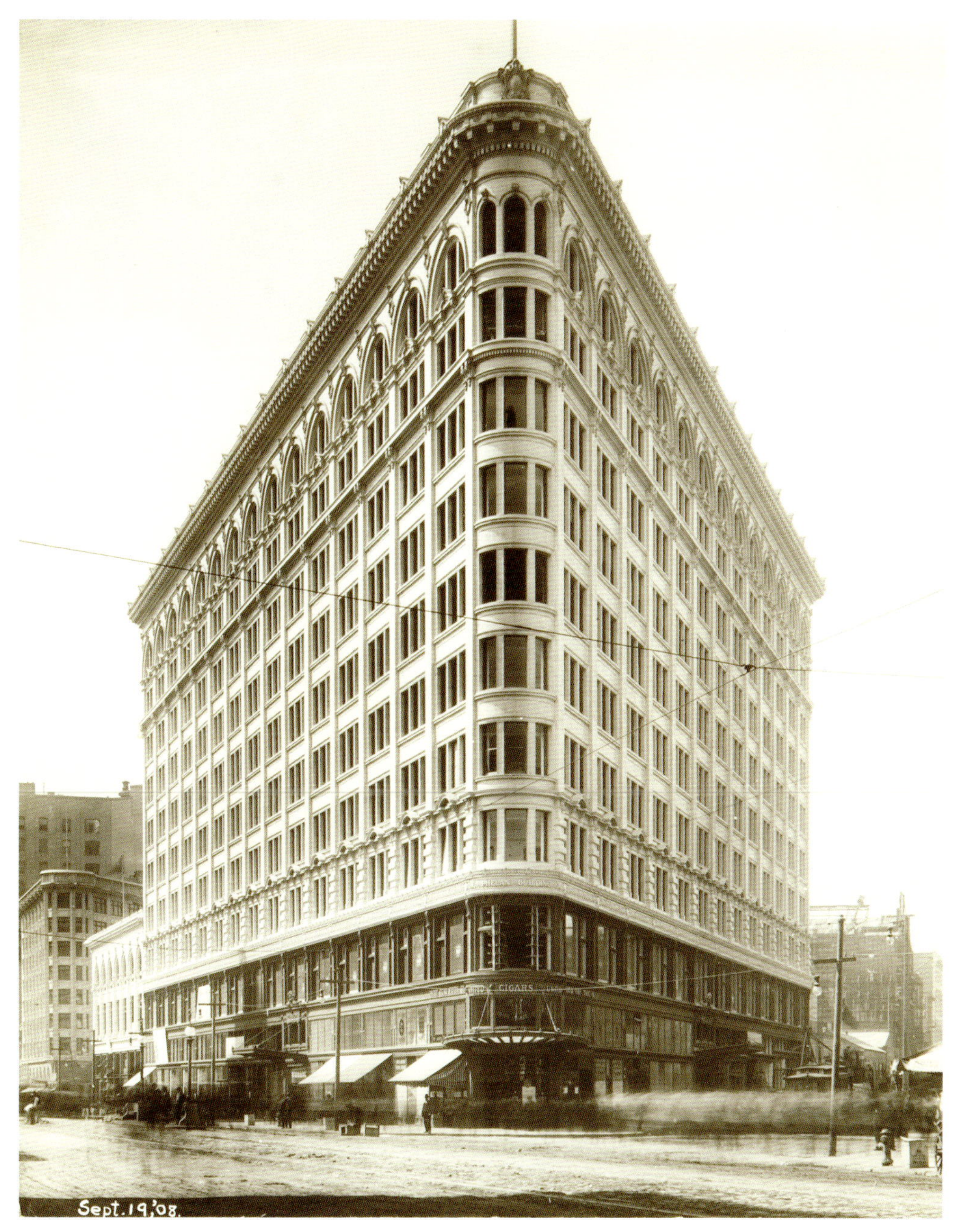

Sept. 19,'08.

PLATE 56. Crittenden Van Wyck, *Untitled,* n.d., four gelatin silver prints
(San Francisco History Center, San Francisco Public Library)

On the upper right of this group is amateur cameraman Van Wyck's photograph of his family's fashionable Eastlake-style Victorian home at 1944 Webster Street in the city's Western Addition. The figures in formal dress pictured coming from the house to the roadster may well be members of his family. In the upper left snapshot is pictured concrete being poured for a new structure at Pacific and Front Streets near the Embarcadero. At the lower left, construction men are shown working on the framing of a new building at Market and Main. And at the lower right, a photograph pictures new trolley car tracks being laid to the east from around Third Street to the Ferry Building. Though telephone poles and lines are visible to the right, much of lower Market still appears to be in varying states of disrepair.

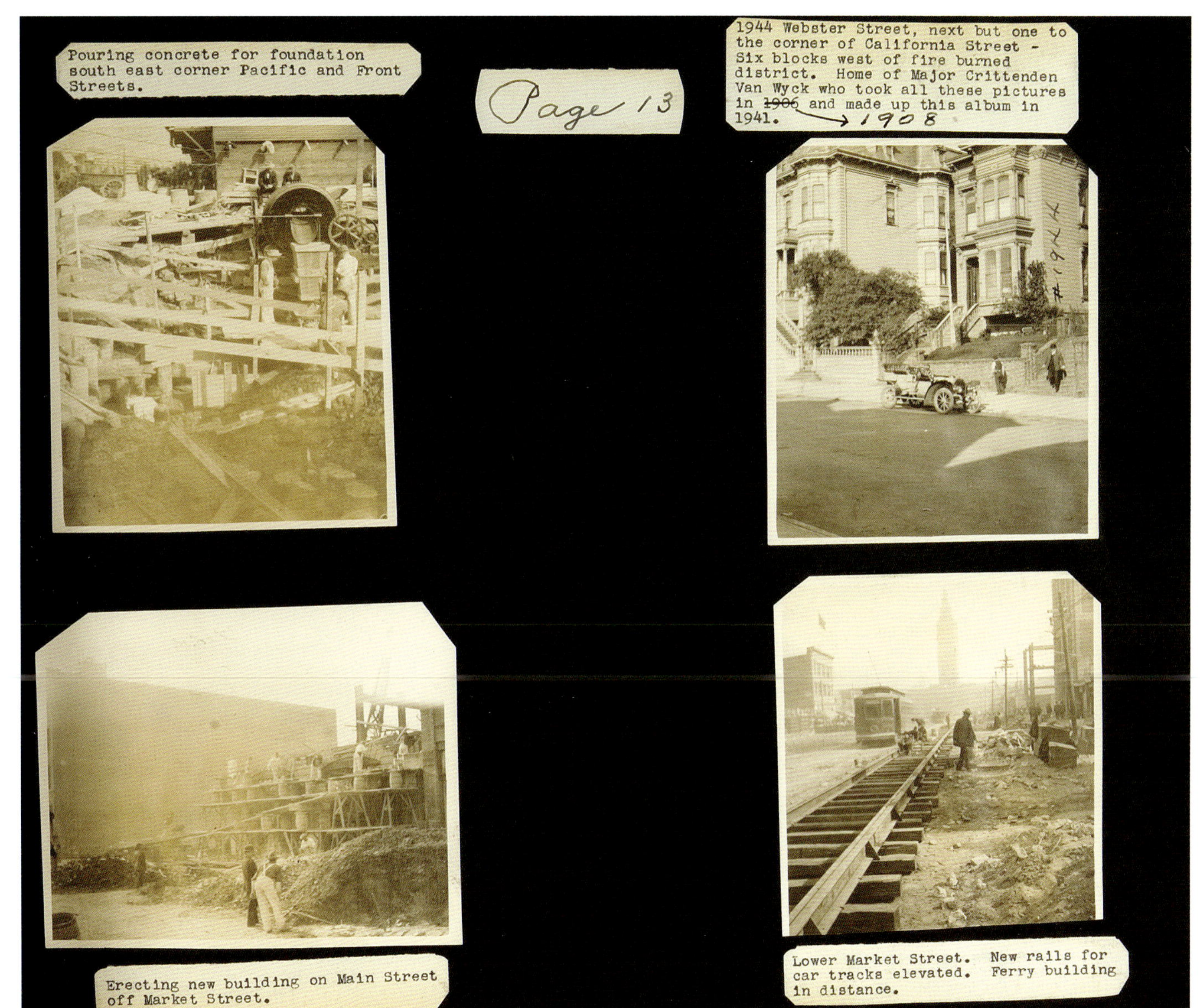

Pouring concrete for foundation
south east corner Pacific and Front
Streets.

Page 13

1944 Webster Street, next but one to
the corner of California Street -
Six blocks west of fire burned
district. Home of Major Crittenden
Van Wyck who took all these pictures
in 1906 and made up this album in
1941.   → 1908

Erecting new building on Main Street
off Market Street.

Lower Market Street. New rails for
car tracks elevated. Ferry building
in distance.

In this group of photos Van Wyck gives us four stages of the city's reconstruction efforts. At the lower left, a scene at California and Battery Streets depicts a site still in ruins at which, the caption tells us, no union men are present because it is a Saturday. On the upper right a horse-drawn dray stands at the location of an excavation made for future construction in Chinatown on the lower slopes of Nob Hill. The walls shown around the site are slated for demolition. At the lower right is the steel frame of the Scroth Building at 240 Stockton on Maiden Lane. This structure, designed by the firm of Cunningham and Politeo, stands to the east of Union Square and was designated as an office building for doctors and dentists. The white X on the photo marks the office that would soon be occupied by photographer Van Wyck who was a practicing dentist in 1908. Originally meant to be an eighteen-story structure, the Scroth Building reached only ten floors due to city height limits mandated after the earthquake. This view provides yet another example of the new steel-reinforced, concrete skyscraper design so universally adopted in the downtown area after the disaster of 1906. On the upper left is the nearly completed West Bank Building designed by architect C. F. Whittlesey. Scaffolding still covers the two-story arcade at the lower level. Another steel-reinforced structure, the West Bank Building, now remodeled, still stands at the corner of Ellis and Market Streets. While the tripartite window design is clear from this snapshot, the triangular or flatiron shape of the building is not. Such triangular form in buildings like this one or the Phelan, Crocker, and Flood Buildings was made necessary by the effect of the diagonally biased Market Street on streets to the north.

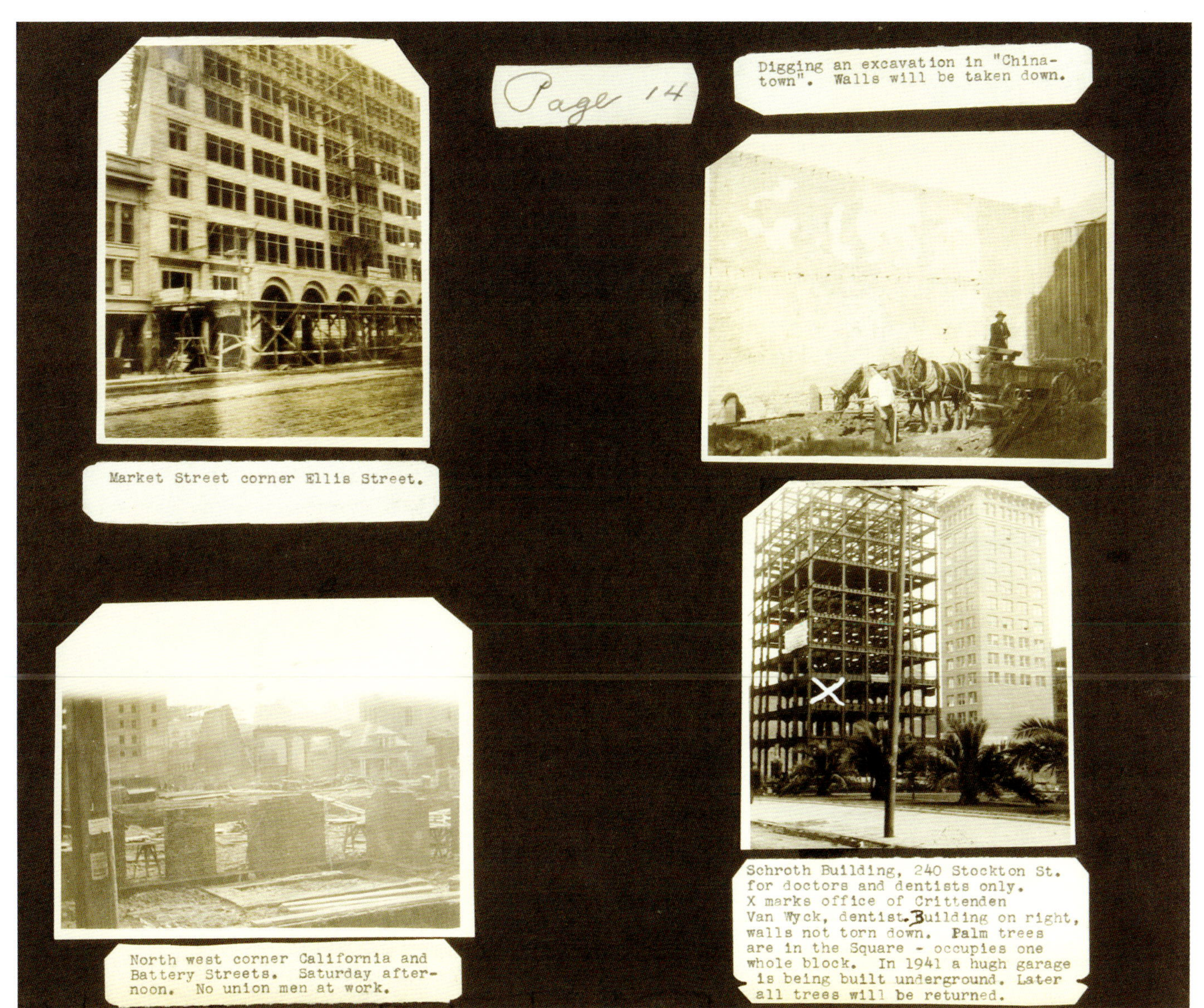

Page 14

Market Street corner Ellis Street.

North west corner California and
Battery Streets. Saturday afternoon. No union men at work.

Schroth Building, 240 Stockton St.
for doctors and dentists only.
X marks office of Crittenden
Van Wyck, dentist. Building on right,
walls not torn down. Palm trees
are in the Square - occupies one
whole block. In 1941 a hugh garage
is being built underground. Later
all trees will be returned.

PLATE 58. Unattributed, *Sutro Grammar School,* 1909, photogravure
(San Francisco History Center, San Francisco Public Library)

In the four years after the earthquake and fire San Francisco voters approved $5 million in bonds for new construction and repair of public schools. This photograph from "Public Work of San Francisco" illustrates how $106,000 of these bonds were spent. Sutro Grammar School, named after former Mayor Adolph Sutro, was located in the Inner Richmond District on Twelfth Avenue. Though under construction here, the brick-faced structure already shows signs of its Edwardian design with classical roofline and flattened pediments surmounting arched main portals. The widespread use of classical design in a variety of San Francisco's major postearthquake buildings, both public and private, suggests the important role that architecture plays in symbolizing the continuity of civilization to a community whose history has been ruptured by cataclysmic destruction.

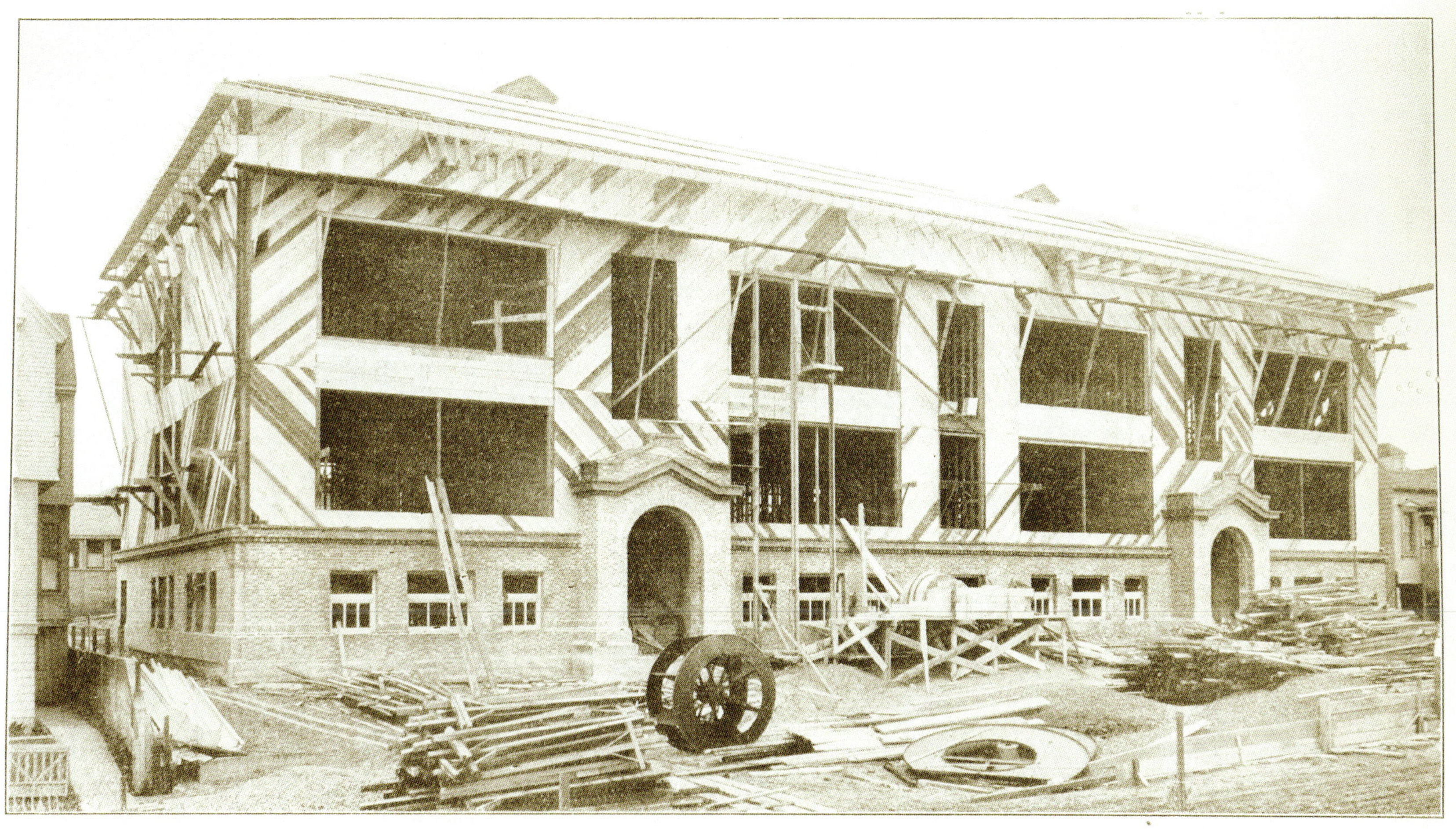

Sutro Grammar School, Twelfth Avenue, Between California and Clement Streets; to Accommodate 720 Pupils and Will Cost $106,000.

PLATE 59. Unattributed, *Bryant Cosmopolitan School,* 1909, photogravure
(San Francisco History Center, San Francisco Public Library)

This is a view of the Bryant Cosmopolitan School under construction between Bryant and York Streets in the Potrero Hill area. Through the heavy scaffolding can be seen such Edwardian elements as arched windows and a classical entrance and roofline. By the time this photograph was published in 1909 the vast majority of San Francisco's 250,000 displaced residents had moved back to their neighborhoods, and well-constructed, pride-inspiring new schools were necessary to serve returning children.

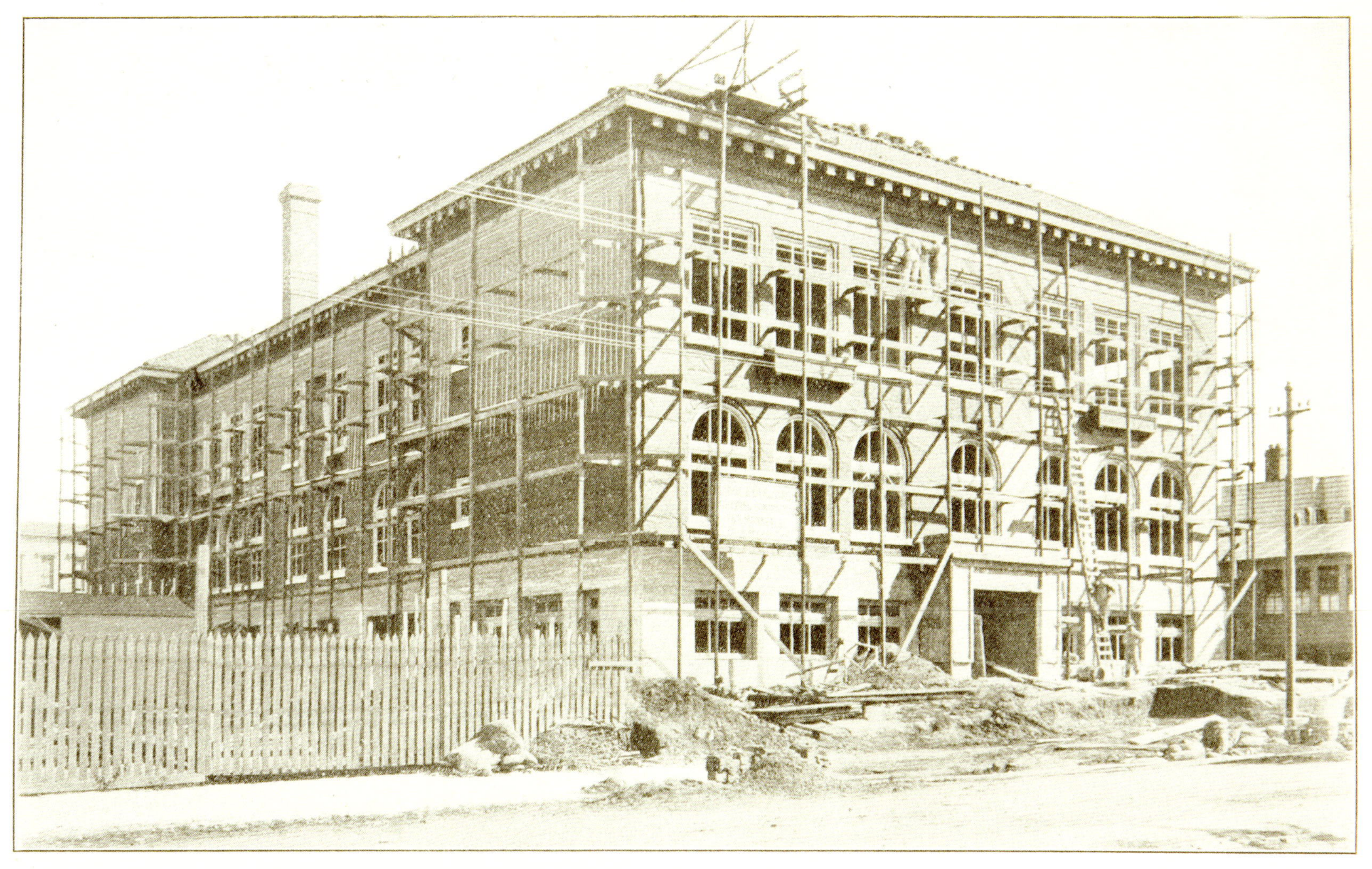

Bryant Cosmopolitan School, Bryant and York Streets.   Will Accommodate 720 Pupils and Cost $110,000.

PLATE 60. Unattributed, *Bergerot School,* 1909, photogravure
(San Francisco History Center, San Francisco Public Library)

The alternating use of brick and stone on the exterior provides much of the aesthetic appeal of Bergerot School built on Twenty-fifth Avenue near Lake Street in the Outer Richmond District. This edifice exhibits great restraint and a good deal of subtlety in its design, not only in its alternation of materials but also in its understated cornice, the delicate scrolls on the uprights of the main floor window frames, the shallow sunburst motif on the front wall of the stairway, and the elegant porthole windows flanking the economically rendered classical entrance to the building. Though the area of the city where this school was located was not devastated by earthquake or destroyed by fire, the fact that San Francisco's government was willing to spend $50,000 on a public facility here indicates that the previously underpopulated westernmost sections of town were now increasingly being seen as integral parts of the reborn city. This new school was named for Pierre Alexander Bergerot, a leading member of the local French community and prominent attorney who in 1898, at the age of thirty-one, became one of the youngest presidents of the Board of Education in the city's history.

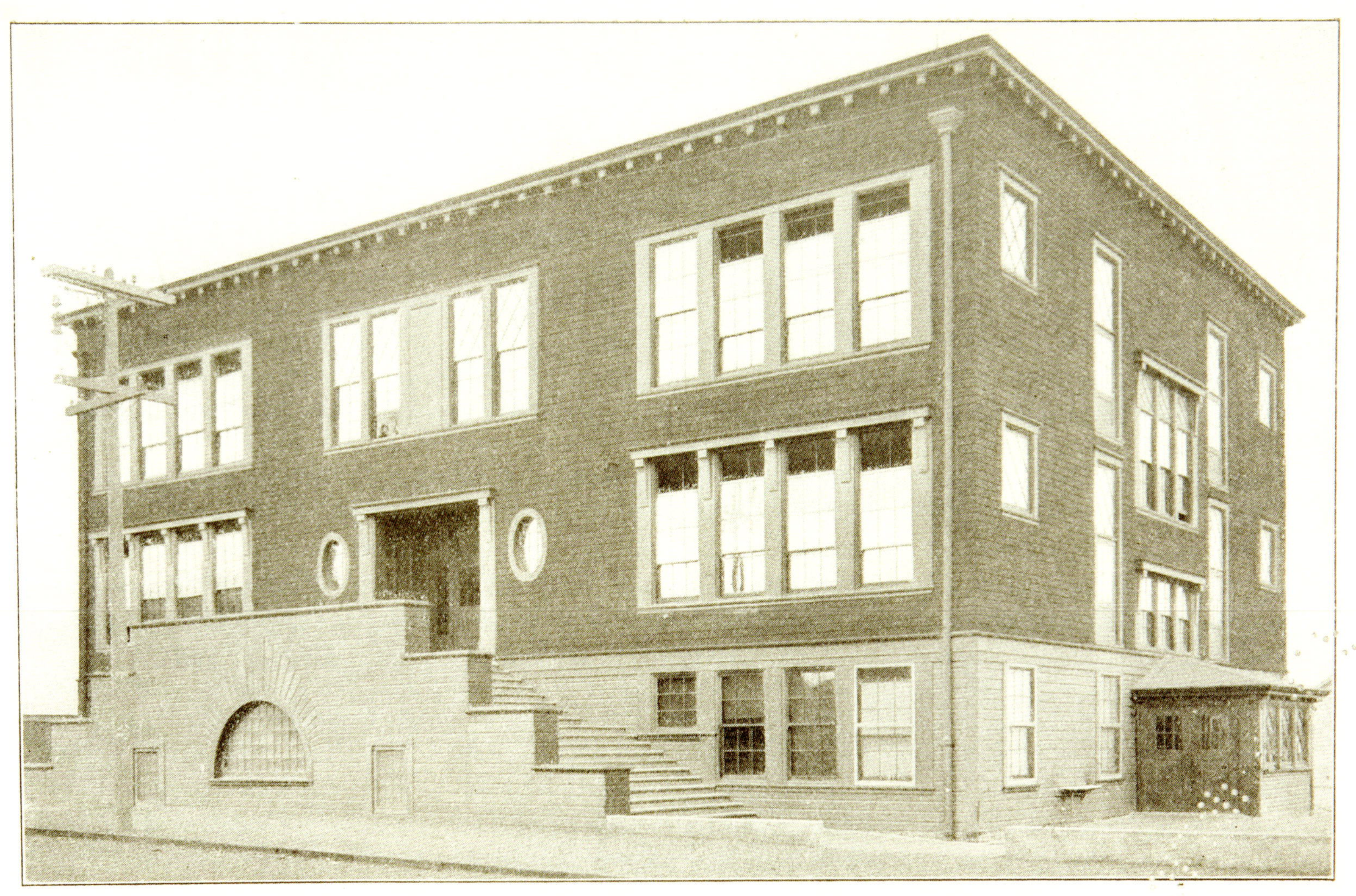

Bergerot School, Twenty-fifth Avenue, near Lake Street. Cost $50,000.

PLATE 61. Unattributed, *Quarters of Engine Company No. 1,* 1909, photogravure
(San Francisco History Center, San Francisco Public Library)

Shown here is the firehouse of Engine Company Number 1 located on Pacific Street near Sansome in the area between downtown and North Beach. This elegant structure provides another version of Edwardian style. Like the Company Number 10 house (pl. 62) it too possesses a wide transomed archway to permit large vehicles access. But this arch is surrounded by a façade decorated in rusticated geometrical patterns. The second level changes style rather radically in its use of such typical Beaux Arts elements as an arched roof parapet with a crest containing the interlocking letters "SF," a heavy cornice line, dentils, circular frieze motifs, inscribed window hoods and corner piers, and two Doric columns in the center. There is some irony in this image since we see a building of traditional design housing increasingly modern mechanized equipment and powered by the new technology of electricity. The six firefighters who pose in front of their new headquarters represent a postearthquake department, which, under then Fire Chief Patrick H. Shaughnessy and his immediate successors, established itself as one of the best in the United States.

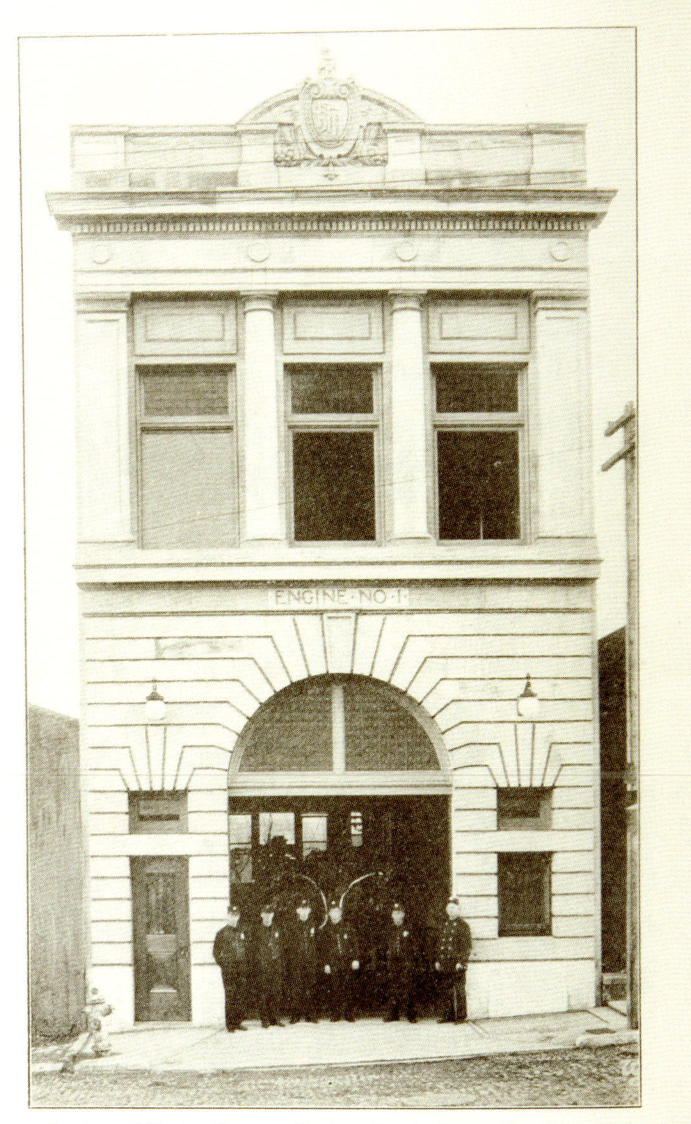

Quarters of Engine Company No. 1 at Pacific Street, Near Sansome.

In the years after the great fire San Francisco did much to prevent the failings that had occurred during the disaster of 1906. A high-pressure water system was put into service and huge water reservoirs were strategically located throughout the city. The fire department that had been severely tested adopted reforms, and several fire stations that had been destroyed were rebuilt. Pictured here is the new building on Sacramento Street housing Truck Company Number 10 with its engine dimly visible through the large entrance. The brick and stone structure was designed in what is generally referred to as the Edwardian style, which is a combination of classical idioms including Greek, Roman, and Renaissance frequently used in the first decade or so of the twentieth century. The two-story first floor that contains the firefighting equipment is defined by a large transomed archway decorated in a somewhat Georgian manner by a stone crest and overlapping irregular stone flange work. A shallow cornice above four medallions separates the two levels of the structure. On the façade of the upper floor meeting room there are three heavily framed windows surmounted by a classical combination of decorative roof parapet, heavy cornice, and supporting brackets and dentils. The architectural care taken on such a small building indicates that public works were still a source of civic pride in the early part of the last century.

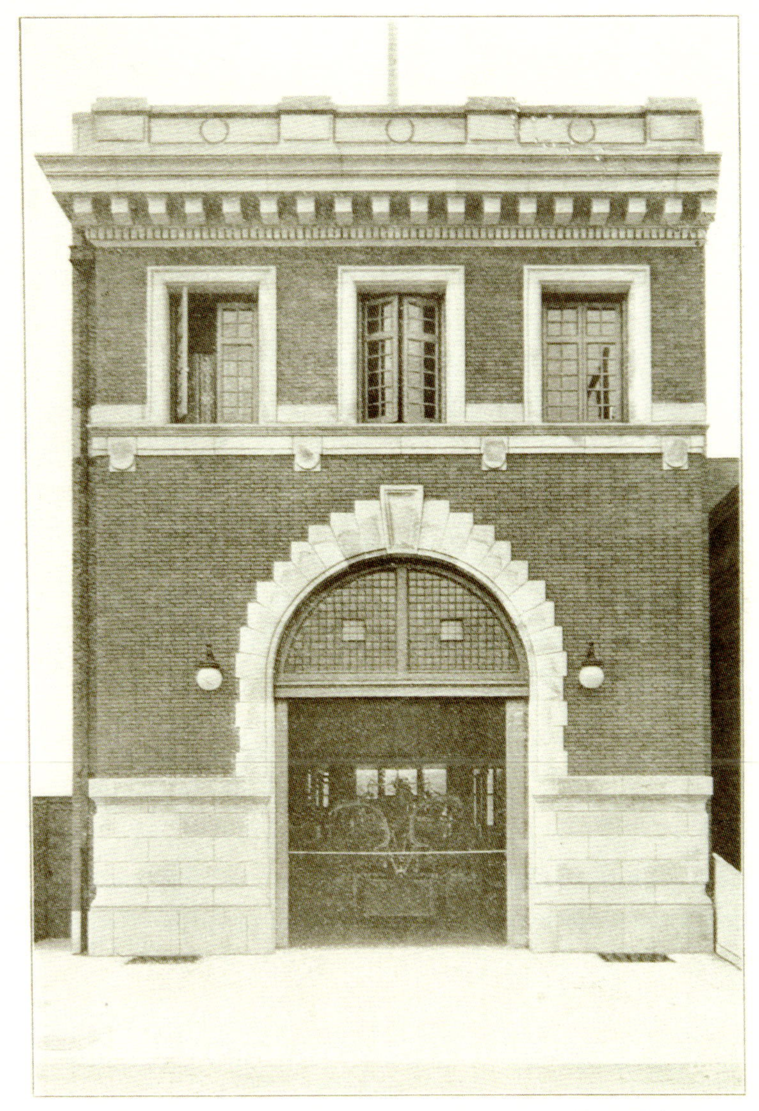

New Home of Truck No. 10, Sacramento Street, Near Walnut.

PLATES 63 AND 64. Unattributed, *Six Feet Circular Sewer in Laguna Street* and *Showing Manner of Reinforcement,*
1909, photogravure (San Francisco History Center, San Francisco Public Library)

One of the greatest tasks confronting the rebuilders of San Francisco after 1906 was the reconstruction and improvement of the city's badly damaged water and sewer systems. A 1907 Department of Interior report outlining the damage done to the urban infrastructure by the earthquake stated the following: "The necessity for firm foundations for sewers running through made ground is clearly indicated. The need of rapid repairs to the sewers is not quite so great as in the case of fire mains, because a city can get along with inadequate sewage facilities, if necessary. It would seem desirable, however, that all important sewers passing through made ground should be constructed of the heaviest iron or steel pipes, and be provided with an adequate foundation." In the years after the disaster $4 million in city bonds were approved for work on the sewer system, and in these two photographs from the publication "Public Work of San Francisco Since 1906" we see that such work was in full swing. The first view illustrates the heavy metal reinforcement structure being put in place to support a six-foot diameter sewer conduit under Mariposa Street in the Potrero Hill area. The second view shows the construction of a six-foot circular conduit running under Laguna Street in the Western Addition of the city. The base of the conduit, looking like an upside down version of an ancient Roman barrel vault, is made of vitrified or surface-glazed brick, piles of which can be seen throughout the site. The Interior Department's call for "adequate foundation" in sewer construction seems to have been taken seriously by local builders, since many of the sewer lines laid beneath San Francisco streets in the flurry of construction activity after the 1906 disaster are still in service over a century later.

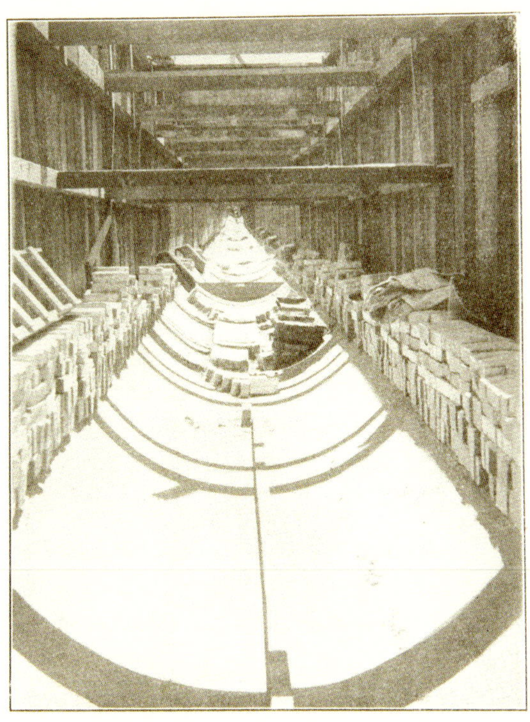

Six Feet Circular Sewer in Laguna Street, Showing Vitrified Brick Invert Lining.

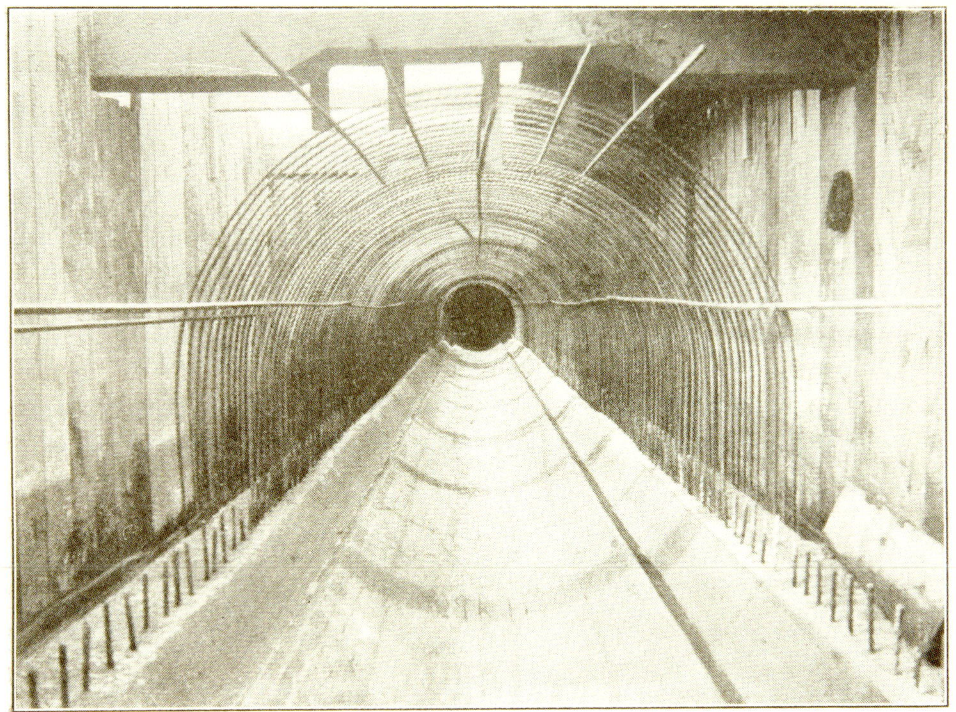

Showing Manner of Reinforcement of 6-foot Sewer in Mariposa Street.

PLATE 65. Willard E. Worden, *Portals of the Past,* 1909, gelatin silver print (private collection)

On the shores of Lake Lloyd in Golden Gate Park stands what historian Randolph Delehanty terms "San Francisco's only architectural monument to the catastrophe of 1906" (Randolph Delehanty, *The Ultimate Guide San Francisco,* [San Francisco: Chronicle Books, 1989], 270). Worden's powerful photograph of this same columned entryway amid the ruins of railroad tycoon A. N. Towne's Nob Hill mansion in the days soon after the disaster became an archetypal expression of civic loss. In 1909, former Mayor and future United States Senator James Duval Phelan arranged for the portico to be moved to the park where it might remain a visible reminder of San Francisco's past, safe from the hands of wrecking crews or the ravages of time. And so three years after creating the compelling image he called "Portals of the Past" (pl. 27), Worden returned to photograph the fragment once again, this time not amid the devastation of recent, overwhelming disaster but rather in a restful naturalistic setting where the mirror image of the "Portals" in the lake suggests that through the passage of time and through quiet reflection the city has come to terms with the horrors of the recent past and can move forward without forgetting them. Indeed, by 1909 San Francisco was largely rebuilt, and the earthquake and fire were already becoming part of the young city's mythology.

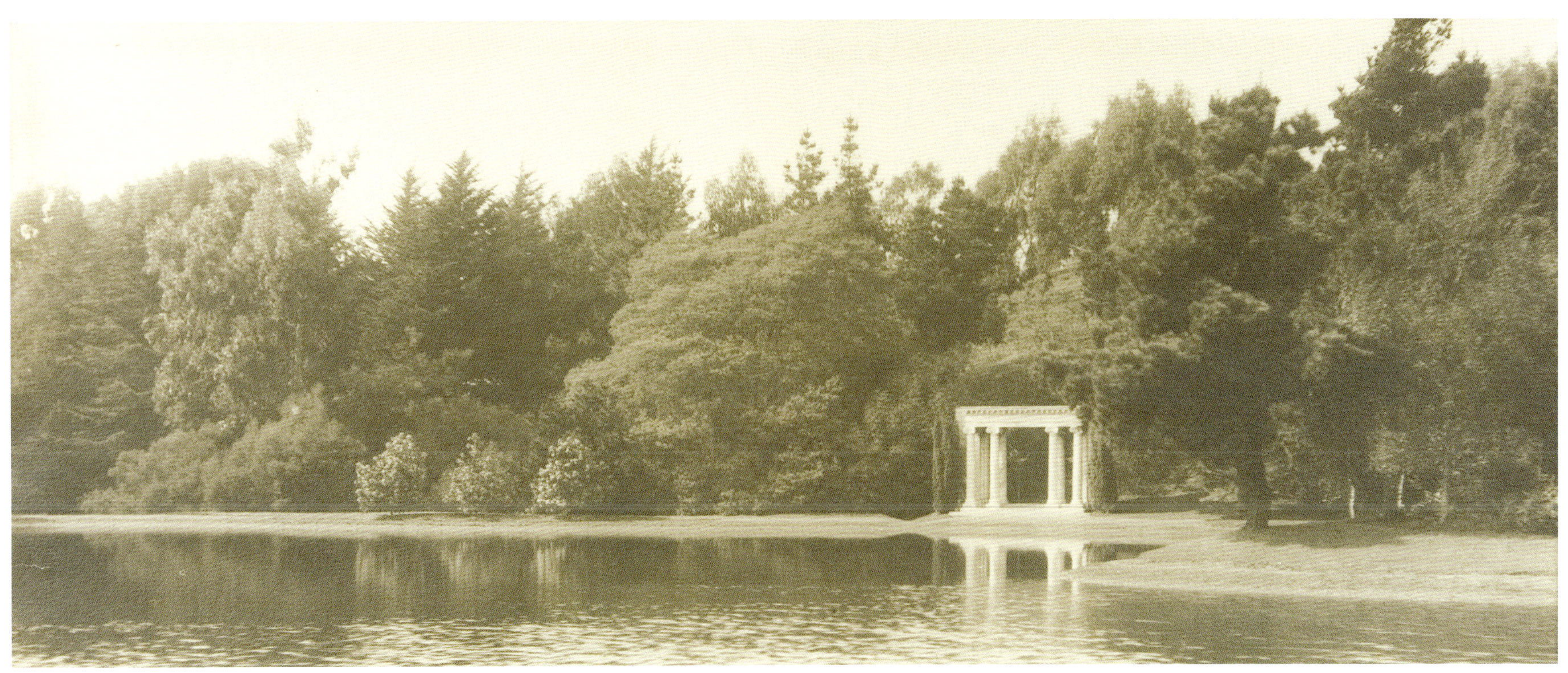

AFTERWORD

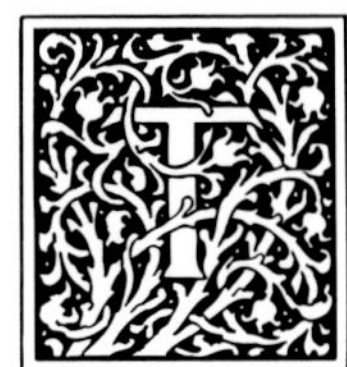

he final image in this book, Willard E. Worden's *Portals of the Past*, reflects much of the instant nostalgia which the disaster of 1906 evoked in the minds of San Franciscans. Images such as Worden's had their literary counterparts in works like Lawrence W. Harris's poem, *The Damndest Finest Ruins*, published in the year of the earthquake and fire, and Gelett Burgess's fascinating novel, *The Heart Line*, which appeared in 1907. Even today, San Franciscans annually memorialize April 18th at a gathering held on Market Street marking 5:12 a.m., the precise moment when the earthquake struck. However, these romantic musings on a world now lost must be squared with the tremendous energy locals exerted in reinventing their city and, in the process, propelling it into twentieth century modernity.

San Francisco's past has been punctuated by decades in which the citizenry—often in the face of economic and political crisis—had shown extraordinary vigor in bringing change to their community. The 1850s witnessed the transformation of San Francisco from a gold rush boomtown of shacks, tents, saloons and muddy streets to a true city of Greek Revival buildings, paved thoroughfares, parks, and a thriving business community with a genuine future. In the 1870s, despite the great national panics of that decade, the local citizenry built the world-renowned Palace and Baldwin Hotels, opened two major music venues (the opulent Wade's Grand Opera and the more democratic Tivoli Gardens), launched the first cable car system in the world, founded the Pacific Stock Exchange and the San Francisco Art Association, transformed Nob Hill with great mansions, opened up the Richmond District to transportation and housing, undertook Golden Gate Park, and began the construction of a new City Hall. During the Great Depression of the 1930s San Franciscans built the Golden Gate Bridge and had a large role in the construction of the San Francisco–Oakland Bay Bridge. Coit Tower was opened, the port underwent revolutionary change, the city-owned Opera House and Veterans' Building were completed, and a world's fair was staged in the middle of the Bay on the newly-created Treasure Island.

Even in the context of these other industrious periods, the decade of 1906 to 1917—when the United States entered World War I—seems unparalleled in the degree to which the local citizenry reinvented a city so recently devastated and shocked by a wholly unexpected cataclysm. While many of the photographs in this book depict the flurry of activity from 1906 to 1909, the task of recreating a largely ruined city continued on with unabated force well beyond this three-year period. The list of accomplishments is stunning: Five hundred blocks of the most densely built and populated parts of San Francisco were rebuilt by 1909 with some 20,000 new structures erected and the downtown area greatly modernized architecturally. The water, sewer, and electrical systems were repaired and expanded. The Fire Department was reorganized, large underground water cisterns were located strategically around the city, and the first fireboats were introduced in 1909. In 1907 a series of political graft trials were held which addressed problems of widespread corruption in the pre-earthquake city government. The Portola Festival, a weeklong celebration of the rebuilt city, took place in 1909. In the same year, the construction of the new San Francisco General Hospital was begun. Two years later the San Francisco Symphony Orchestra was founded and in 1912 the Municipal Street Railroad began operation. An elaborate Civic Center was laid out in 1912, and by 1914 John Galen Howard's Exposition Auditorium opened with Bakewell and Brown's City Hall being dedicated in 1915. Another great civic project, the Hetch Hetchy water system, was undertaken in 1913, and the Stockton Street Tunnel connecting North Beach to the downtown area was completed in 1914. In 1915 San Franciscans staged one of the greatest of all American world's fairs, the Panama Pacific International Exposition, an event attended by nearly 19 million people and covering 635 acres along the Bay shore in the Marina District. In 1917 the Twin Peaks Tunnel connecting the southwestern section of San Francisco with the downtown by rail line was opened; a grand new neoclassically designed Public Library took its place in the Civic Center; and architect Willis Polk created the Halliday Building on Sutter Street for the University of California, the first completely glass-fronted skyscraper built in the United States.

And so, along with the memories of a city lost and the yearnings for less complicated times, the disaster of 1906 initiated an era of energy, enthusiasm, boosterism, pride and forward-looking civic effort unrivaled in San Francisco's dramatic history. Indeed, many of the achievements of that historical moment remain intrinsic parts of San Francisco's physical presence and cultural life at the beginning of the twenty-first century.

Marvin Nathan